Flow

The Science of Fluids
and
Timeless Lessons They Leave Behind

T.J. Khan

InkTangible
From Thoughts to Words

This book is a work of nonfiction. All scientific information is accurate to the best of the author's knowledge at the time of writing. Some scenes include narrative elements for the sake of immersion. While many of these stories are drawn from real events, some have been creatively imagined to serve as thoughtful entry points into the scientific ideas that follow. The author and publisher disclaim any liability arising from the use of the material in this book.

ISBN: 979-8-9994688-8-8 (paperback)
ISBN: 979-8-9994688-3-3 (ebook)
First edition

Designed & published by **INKTANGIBLE**

We'd love to hear from you: contact@inktangible.com
You can also visit us at: www.InkTangible.com

InkTangible
From Thoughts to Words

Dedication

To My Parents

Who provided for me as a child,
And continue to do so whenever I need, even as I raise my own.

Contents

What's Inside

0 — The Buildup

What is Fluid Mechanics, and why should we care? *Hint:* It is not about equations, and it is not for scientists. It is about the everyday world, and for everyone living in it. It flows under our noses, through our veins, and in our lungs, isn't it time we let it into our heads?

1 — Visible Yet Unseen

Why do we always find a Kenyan marathon runner on the winning stage? Also they say a watched pot never boils—I say, try that in the hills. There is a force we can't see, but is visible all around us since we feel it every second of every day. It even affects the car tires and our beloved snack packs.

2 — A Deep Dive

What happens under water? From swim bladders in fish to the perils of diving. The deepest scuba dive ever recorded took 15 minutes during descend, but 13 hours to return. We'll get to know the whys. Is pressure a villain or a superhero? It all depends on who's telling the story.

3 — The River Within

Every heartbeat sends a pressure wave that travels through 60,000 miles of vessels. What is millimeters of mercury, and what does that has to do with our blood pressure? The flow of blood has a timeless advice for us: keep moving, no matter what.

4 — Twist in the Skies

Why does air move at all? In one part of the world, winds blow endlessly in one direction only; in another, they entirely vanish and in some the direction is reversed. What happens if a pilot forgets the Earth is spinning beneath them? Well, I landed in Kiribati instead of Ghana. *Exactly...* I didn't know where Kiribati was, either.

5 — And a Storm is Born

Why do storms form, and how do they spin? In fact, do they really spin or is that just about our perception? And if it is all a matter of viewpoint, what about the storms we conjure in our heads against others? Is the real danger in the winds or in how we insist on seeing everything from our own vantage point?

6 — The Story of Flight

For humans flying was a dream until we found the *WRIGHT* way of doing it. We'll see all the details. Before taking off, I'll introduce you to my friend Bernoulli. Harmless guy. The only trouble is, once you meet him, he starts following you everywhere. I recently spotted him in a showerhead. *Yeah... I know...* He's a gentleman, though. Always knocks on the head before entering your brain.

7 — To Swing or Not to Swing

Why do golf balls have dimples? Well, if you're being smacked with a club and flung skyward, at least go with a pretty smile. I promise we'll get to the real reason. By the way, how does a cricket ball swing through the air, then out of nowhere change the direction of swing just when the batters think they have figured it out? Physics and bowlers clearly share a wicked sense of humor.

8 — The Art and the Artist

Hundreds of thousands of goals have been scored in soccer, but only a few will be remembered as long as humans walk the Earth. Roberto Carlos' free kick is one of them. It was a piece of art painted in thin air. The brushstrokes were physics, and the palette, fluid mechanics. We'll also see how tennis balls dip and how pitchers in baseball make fools out of batters.

9 — The Poetry of Fluids

In 2006, a painting was sold for *140 million dollars*—the highest price ever paid for a work of art at the time. What made it special? It captured the poetry of fluids in its purest form. No brush imposed control; the fluids spoke in their own language. We'll see the role of fluids in the elegant world of art. Also isn't it fascinating that the same forces that pull water up through a plant's roots also breathe soul into a Van Gogh painting?

10 — The Architect of Modern Living

With a mere flip of a switch, we get cool air, flowing water, and steady light. It wasn't always this way for our ancestors.

Life got smoother when fluid mechanics rolled up its sleeves. From pumps that push to turbines that power, this chapter peeks inside the machines that keep our world moving. Oh, and how could we forget what powers our mind? We'll spill those beans too...

11 — A Tale of Betrayal

A drop of water ascends to the skies only to fall back down. It travels across lands, shapes valleys, carves mountainsides, and finally returns to the ocean... right where it began. Isn't this journey symbolic of a larger truth? We will also witness a story of betrayal where a river changed its mind. It left like a lover who suddenly grows distant and walks away, leaving behind no explanations—only questions, the scars of its absence, and the lingering ache of abandonment.

12 — Flow of Life

Just like fluids, life flows too. Unlike rivers or winds however, we are often chasing, rushing, grasping, unsure of what we are truly seeking. In that blind current, chaos begins. To understand life, we must understand its fluid nature. At the end of the day, we are a living expression of fluid mechanics.

0

The Buildup

It was a rain-soaked afternoon. The café was as cozy as ever, with warm lights and soft music playing in the background. Wrapped in a woolen sweater, coffee in hand, I was listening to the rain on the other side of the foggy windows, completely lost in the tranquility I keep seeking amid the noise of the world.

Outside, however, behind the curtain of rain, something greater was happening. Something that stretched beyond the definition of tranquility. Streams of water were running down the street, slipping through cracks and drains, finding their paths in what appeared ordinary... until one realizes that this flow has been shaping the Earth for billions of years. Way before you and I ever had our first sip of coffee, or back further, when neither we nor coffee were even a mentionable thing.

And rain is just one part in the larger story of flow that drives the oceans, drifts the clouds, and spins the storms. That pumps blood through our veins, gives life to the art we admire, and yes, bends a football into the goal in ways that seem impossible.

This story of flow is the story of **fluid mechanics**.

Until now, you may have known it as just another branch of physics. Hard, technical, complicated, or so they say. Yeah, sure! All these, if one is sitting in a room lit like a morgue, in front of someone who thinks excitement is an unprofessional thing and enthusiasm a health hazard. And all one can see are equations filled with symbols that could easily be mistaken for scary snakes or confused hieroglyphs. Honestly, if someone tells us how to drink tea with this approach, using formulas and all, it'd feel painfully technical too, wouldn't it?* Imagine using vectors to find the right slurping angle, or a differential equation to calculate biscuit dunk time, or accurate speed of air blown from the mouth to remove heat. Enjoy the tea, Sir!

The point is, this isn't true. And when it comes to fluid mechanics, it couldn't be further from the truth either. Because, like drinking tea, it is neither hard, nor technical. It's just that no one ever really shared it. And when they did, they shared it in exactly the way we agreed it shouldn't be: dry, dull, and needlessly boring.

What it really is, is not something distant or abstract but the world we inhabit and the beautiful reality we live every day. Remember *that* stunning free kick by Roberto

---

*By the way, I don't mean tea made with formula milk. That would certainly be a bad experience.

Carlos against France? That was fluid mechanics in its full glory. Or if you are a cricket aficionado, how about Wasim Akram's reverse swing deliveries. Well, same story. Even something as simple as a tiny drop of ink swirling in a glass of water, creating mesmerizing patterns, shows its elegance. It is everywhere around us.

So, what exactly is fluid mechanics? Anything that can flow is a *fluid* and this includes all the liquids and gases, like water and air. This is the reason why we call someone "fluent" when they speak smoothly, it's as if the words flow from their mouth like water from a stream. *Mechanics*, meanwhile, is how things act under the influence of forces. Combine the two, and fluid mechanics is simply how liquids and gases behave in the presence of forces and how they interact with the solids around them.

Things that move are *dynamic*; those that remain still are *static* (just like personalities). Fluids, too, can either be in motion or at rest. When they are moving, we call it fluid dynamics, such as wind rushing past a sail or water flowing through a pipe. When fluids are at rest, like water in a still pond, we refer to it as fluid statics. Together, they make the two pillars of fluid mechanics.

I totally understand that somewhere in your mind, you must be thinking, "But why should I even care about *fluids* and their *mechanics*?" Let me show you why. The air we breathe, sustaining us with every breath, is a fluid. The water we drink, keeping us hydrated with every drop, is a fluid. And inside us, blood, a fluid like no other, flows to deliver oxygen and nutrients. To cut to the chase, we are surrounded by fluids—inside and out. They're the lines that sketch us, and without them, life as we know it wouldn't exist.

So once we begin to notice *how* fluids flow or rather *why* they flow the way they do, the world starts to shift right before our eyes. We no longer see rain as isolated drops but as the origins of rivers and oceans, and that football bending into the top corner turns into live magic. Fascinatingly, understanding this magic will not spoil the mystery for us. It makes it even more astonishing, showing layers of magnificence we never knew existed.

This book is your invitation to see the spellbinding world of fluids. And **NO**, it is not a textbook. In fact it is like you have just walked into the same café where I was sitting earlier. We've settled into a corner with warm drinks in hand, and now we are talking about the fluids that how amazingly alive everything really is when we look a little closer. You won't need a PhD, a degree in physics, or a degree in any other subject to enjoy its beauty. If you can read—and clearly, you can—you are already qualified. My goal is to make it so straightforward and so vivid that anyone, of any background or age, can easily follow along. Einstein once said,

> *Things should be made as simple as possible, but not simpler.*

We will follow his advice and keep it simple yet accurate, skip the unnecessary jargon, and focus on what matters the most. As fluid mechanics isn't about equations and formulas; it is a collection of some worth telling tales that aren't just reserved for engineers or physicists but belong to everyone, because, whether we realize it or not, we're all part of them. From rain in the skies to breath in our lungs are all chapters in this grand tale.

And just when we think this is purely about science, it nudges us to think again. There's more to it than meets the eye. Besides explaining how fluids flow, the principles of fluid mechanics offer insight into how *we* should flow through life. They carry timeless truths—worldly wisdom, if you will—that help illuminate our path.

If this book does anything, I hope it helps you see the ordinary moments through a different lens. You might find nature speaking, sharing its captivating secrets and offering lessons that were always there, just waiting for you to listen.

Ready?

Let's FLOW...

1

Visible Yet Unseen

Every year, on the third Monday of April, thousands of runners from all over the world gather at the starting line of the Boston Marathon. The dreams in their hearts are asked to stretch across 26.2 miles. With a history of more than a century, it is one of the most celebrated races on the planet. Global estimates suggest only a sliver of humanity runs a marathon in any given year, roughly 0.013 to 0.016 percent of the world's population. Fewer still cross that line more than once. The people who show up here are the outliers among outliers.

Yet year after year, in Boston and across the marathon world, one familiar scene repeats itself. When the medals are handed out, a Kenyan athlete is almost always on the podium, as if it is the ritual of the race. The whispers that follow are just as regular. It's their mentality... no, physiology... could be genetics. For some, however, the elusive secret is believed to be their training ground. A remote, unassuming place that few outside the running world could name, but one that has produced a surprising share of the best distance runners—Iten, often called the Home of Champions. But why there? What gives that patch of earth such power?

The path to the answer does not begin in the genome. Instead, it is rooted in something so ordinary we forget it exists: an invisible force that touches every breath we take and affects every step we make. It sets ceilings on how high we can climb and puts floors under how long we can endure. It is the *air* and its *pressure*.

Imagine standing on the beaches of Dubai, where the air fills the lungs with ease. A gentle breeze brushes past, and everything feels soothing. Breathing here, at sea level, is as natural as the Arabian Sea itself. Now shift the scene to a hill station, like Shimla in India at 7,200 feet (2,200 meters). Here, each breath takes just a little more effort. Every inhale feels slightly strained and our lungs seem to be working considerably hard. It makes us worried if something has happened to them. Well, the lungs are as capable as ever, it is the air itself that has changed.

Let's dive a bit deeper into the concept—or better yet, into a pool of water first. As soon as we jump into the swimming pool, the weight of the water presses down on us and pressure starts to build on our body. The deeper we go, the more of it stacks above us, increasing the pressure with every inch we descend. The vast atmosphere that surrounds us is basically a gigantic pool, not of water, but of air. And just like water has weight, so does air. Gravity pulls it toward Earth and this ocean of air exerts force, pressing down on everything below it. This is *atmospheric pressure.*

The Earth's atmosphere is a mixture of different gases, mainly nitrogen and oxygen, stretching from the surface of the Earth to the edge of space, about 300 miles (480 kilometers) high. At sea level, atmospheric pressure is at its highest as there's the most air above us, pressing down. However as we ascend, say, while climbing a hill like Shimla (2,200 meters or 1.4 miles), the air pressure decreases since the higher we go, the less air there is. At Shimla, we are breathing in less oxygen with each inhale, and the air feels thinner simply because there's less of it present at that altitude. Even if we have a strong pair of lungs, physical activity or a steep walk feels harder. Similarly, in other high-altitude cities, like Denver which is famously known as the Mile-High City (1,600 meters), many people find themselves short of breath upon arrival.

But wait a minute... why does climbing just 1 or 2 miles above sea level make such a big difference? Doesn't the atmosphere stretch up to 300 mile? At places like Denver or Shimla, there's still about 298 miles of air above us, so shouldn't the pressure hardly change? Well, the thing is that the air isn't spread out along 300 miles *evenly.* Most of it is packed close to the Earth's surface and is concentrated near the ground due to gravity, compressed and squeezed under the weight of the layers above it, making it denser and more pressurized. It can be thought of as a set of clothes in your wardrobe. The ones at the bottom are tightly squeezed under the weight of the rest, while those at the top remain loose and fluffy. Assuming, of course, you're the tidy type. Else your "air layers" would probably be scattered across the bed and the chair and the floor.

So in other words, the change in air pressure isn't gradual. Instead of decreasing evenly along the entire 300-mile-high atmosphere, the drop is much steeper in the lower altitudes. At sea level, like in Dubai, we are surrounded by thick air and the pressure drops by around 3% for every 1,000 feet (about 300 meters) we climb. By the time we reach only 8,000 feet (2.5 kilometers), the air is already about 25% thinner. In fact, around 75% of the atmosphere's total air is packed into the first 7 miles (11 kilometers) above sea level. The elevation of any hill station may seem insignificant compared to the atmosphere's towering 300 miles, it's very much a meaningful climb into thinner air. For those living at high altitudes, breathing in such an environment where pressure is low and oxygen is limited becomes the norm as their bodies adapt over time and breathing feels easier. To visitors, however, atmospheric pressure rather rudely tells not to take it for granted.

Now, what if we turn the scenario on its head? That is, if low-altitude visitors struggle when they ascend, what about the reverse? Does someone living in high-altitude environment notice a difference in breathing pattern upon descending to sea level? To find the answer, let's go to Iten...

The Secret of Mountain Runners

Hidden in the Kenyan highlands, Iten sits at 7,900 feet (2,400 meters) above sea level where the air pressure is low, and oxygen, like a rare commodity, is harder to come by. This makes every breath feel like a fight. But this struggle doesn't go to waste as it pushes the body to adapt. Red blood cells in our bloodstream pick up oxygen from the lungs and carry it to the muscles. If oxygen is in short supply, our bodies gradually start producing more red blood cells to increase the oxygen carrying capacity to cover up for the shortage. The more red blood cells there are, the more oxygen can reach the muscles. This natural acclimatization turns the bodies into endurance machines.

When such athletes descend from high altitudes to race at places like Boston (situated at sea level), where the air is already rich with oxygen, the presence of extra red blood cells in the blood brings a surreal effect. The sheer abundance of air tastes like an indulgent luxury. It's as if a car designed for low-grade fuel is, out of nowhere, fed the best quality high-octane jet fuel. If that doesn't paint the picture well, imagine someone trained to run on crutches, that too chained, and then suddenly the crutches and the chains are removed. With high amount of oxygen reaching the muscles, the pace naturally gets quicker and the stamina becomes deeper.

The world has caught up with this secret and the idea is now widely popular as *live high, train low*. Athletes go to high-altitude locations to gain this edge which forces the body to produce more red blood cells. They come down at low-altitude locations to train themselves so as to maintain the training intensity and when it's time to race at sea level, the body is primed to perform at its peak, pushing faster and farther than before. These adaptations aren't permanent though and the body reverts to its natural state within weeks of returning to lower altitudes. Which is why timing matters so much as successful athletes make sure they hit their spike exactly when it counts—the race day.

Coming back to Iten, on paper, yes it does offer high altitude and difficult terrain. However talk to those who have trained there and they will mention something more which is rather less measurable. The air may be thin in oxygen but it is thick with passion. It is rich in shared spirit of sweat and laughter that no barometer can capture. The feeling of being surrounded by others chasing the same dream, and somehow wanting you to win, too. Athletes are drawn here by this legacy of greatness which, I believe, is its true elevation. This rare combination of physical adaptation and cultural drive is what has produced some of the greatest marathoners the world has ever known. One could argue that endurance might come from the height, but there's no arguing that it is discipline and commitment that shapes the kind of character needed to rise to the top.

Has all this talk about breathing and running left you a little breathless? Let's grab something to eat and head out on a drive to the Cederberg Mountains in South Africa. We start our road trip from Cape Town, a lively city at sea level along South Africa's stunning coast. We stock up for the journey with an assortment of snacks. Some chips, of course—or crisps, if you're feeling a bit royal.

The car winds its way up the mountain roads. The air turns cooler, and with every mile, the scenery becomes more breathtaking. After a couple of hours of driving, we finally reach the destination. You stretch your legs, open the car door, and reach for the pack of chips in the back seat, only to notice something odd...

The pack is puffed up, tight like a balloon, and looks ready to burst at the slightest touch. It looked normal when you picked it up from the store back in Cape Town. "What has happened... Have the chips turned bad?" You ask, surprised.

Down at sea level, when a pack of chips is sealed, the air inside and outside the bag are at the same atmospheric pressure. As we climb higher, the atmospheric pressure drops, but the air pressure inside the bag stays the same as that of the sea level (provided the pack is properly sealed). With the internal pressure now higher than the external, the bag gets expanded and snack pack looks like it's been inflated with a bicycle pump. And yes, if the altitude gets high enough, *Pop!* The pack of chips can burst open. Don't blame the packaging company though, blame physics.

Also it's not just our pack of snack reacting to the changing pressure. The car tires also feel the effects of the journey. Let's dig in a bit as the devil loves to hide in the details. We start with the units in which pressure is usually measured.

At sea level, the air pressure represents the entire weight of the atmosphere above us, so scientists named the unit of pressure as 1 atmosphere (atm). In that meeting, someone must've said, *Let's not confuse people today and just call it what it is.* However, there are a dozen other pressure units waiting to spread confusion, like pascals (Pa), millimeters of mercury (mmHg), pounds per square inch (PSI), bars... and believe me, many more. We use different units depending on the situation. For example, blood pressure is measured in mmHg, weather forecasts often mention atmospheric pressure in millibars, while for car tire pressure, some countries use units like bars while others use PSI. Let's stick with PSI for now.

When we say a tire is inflated to 32 PSI (pounds per square inch), it means that every single square inch of the tire is being pressed on with a force of 32 pounds. To put it plainly, it's like placing a 32-pound weight on every square inch of the tire's surface (a square inch is a one-inch by one-inch square).

Before hitting the road from Cape Town, let's fill the tires with air for a smoother ride. You might assume the tires are completely empty before we add air but that's not quite true. Even a tire that looks completely flat (or punctured) still contains some air, due to the surrounding atmosphere. So when we inflate a flat tire, we are not starting from zero, but adding air on top of the pressure that's already inside the tire—air that's present because the tire is exposed to the atmosphere. In Cape Town (at sea level) this pressure would be 1 atm which is equal to 14.7 PSI.

Let's say you pump in 32 PSI of air. That means the total pressure inside the tire is effectively 46.7 PSI (32 units from the air you added, plus 14.7 units from the atmosphere). The interesting part is when we check the tire pressure with a gauge, it does not show the total pressure (46.7 PSI). Instead it is designed in such a way that it only shows the pressure difference between the inside of the tire and the air outside, that is the extra pressure above the atmosphere. This difference—32 PSI in this case—is called *gauge* pressure, because it simply gauges how much more pressure is in the tire compared to the air around it. If we check a flat tire with a gauge, it shows zero as the pressure inside and outside are the same (atmospheric) and there's no difference to measure. Okay, now what happens when we drive to the mountains and check the pressure again?

As we climb higher, the atmospheric pressure drops from 14.7 PSI or 1 atm (at sea level) to say 12 PSI or 0.8 atm (in the mountains). The total pressure inside the tire stays the same at 46.7 PSI. When you check the pressure with a gauge here, it will show a bigger difference:

46.7 (inside) - 12 (outside) = 34.7 (gauge)

Though you haven't filled any air, the gauge reads higher than what it showed before, tempting you to believe that the tire has somehow gained air through sorcery only found in the hills.

We could say that atmospheric pressure at sea level acts like a strong man giving a solid punch but as we drive higher, that strong man turns into a little child, punching with much less force. The air inside the tire (or the snack packet) doesn't change, but without the strong punch of atmospheric pressure to push back, the tire shows higher gauge pressure, feels firmer, tighter and more pressurized.

Even a Watched Pot May Boil...

We have spent the entire day driving through the rugged beauty of the Cederberg Mountains. By the time we reach the cabin for the night, your stomach is growling, and all you can think of is dinner. How about we cook some pasta?

You fill a pot with water, set it on the stove, and wait for it to boil. The water starts boiling quicker than you expected, way faster than back home at sea level. You toss in the pasta and wait... and wait... and then wait some more. Your patience, like the boiling water, is bubbling over, but the pasta stubbornly refuses to cook.

You always thought that as soon as the water starts boiling, the food cooks quickly. This might come as a surprise but just because water is boiling doesn't mean it is actually hot enough to cook. Yes ! It is not the boiling that matters. What matters is the temperature of the water and the heat, and how efficiently that heat is passed to the food. Let's cook the concept first, shall we?

Boiling is the process in which liquid water turns into vapor or steam. For this, two things need to happen. One: the water molecules need enough energy to break free from the forces holding them together in the liquid form. Two: they also have to push back against the weight of the air pressing down on the surface of the water. Only when the molecules have enough energy to do both—break the bonds and fight off the pressure—can they escape into the air as vapor.

At sea level the thick and heavy air keeps water molecules tightly packed together. The atmospheric pressure acts like a lid, requiring more energy (higher temperature) to push the water molecules into the gas phase. We need to heat it to 100°C (212°F) to make the liquid water escape into the air as steam, which is the boiling point of water at sea level. A useful analogy is to picture water molecules as kids trying to jump on a trampoline. The stronger someone is pressing down on them, the harder they have to jump to break free. But up in the mountains, the air pressure is lower, hence there's less "weight" of air holding those water molecules down. Here, they can escape into steam more easily, and so the water boils at a lower temperature, say 90°C (194°F) instead of 100°C (212°F).

Great as it may seem that the water is boiling but it's not as hot as the boiling water at sea level. What is even more interesting is that once water reaches its boiling point, its temperature stops rising. No matter how much more heat we apply through the stove, it won't get any hotter. All that extra energy goes into turning the remaining water into steam rather than raising its temperature further. So, when water comes to a boil, that is as hot as it can get—100°C (212°F) at sea level and 90°C (194°F) in the mountains.

The pasta doesn't care if the water is boiling or not; it is concerned about how hot the water is. It needs enough heat to cook properly. At sea level, water absorbs more heat from the flame until it boils at 100°C. This breaks down the starches and proteins in pasta, cooking it quickly. In the mountains, when water boils at 90°C, it doesn't absorb enough heat and so there's less of it available to do the cooking job. Since once boiling begins, water can't absorb any extra heat, the same noodles that cook in 10 minutes back home might take 15 or even 20 minutes here.

Now, if you are really hungry and looking to speed things up, you could head to Death Valley in California, which sits at -86 meters (-282 feet) below sea level. There, the atmospheric pressure is slightly higher than normal, so water boils at around 100.1°C (212.2°F), just a touch hotter than at sea level. This tiny boost in temperature means food cooks a bit faster in Death Valley, as more energy is being transferred to the meal with every second.

STOP! Don't go. Upon reflection—which, in my case, always arrives late—I admit that suggesting with great enthusiasm, a trip to a place with a name as ominous as *Death Valley* to a starving stomach already on the verge of collapse, was perhaps not my most luminous idea. It was as bright as recommending a sun bath to a man dying in the Sahara. I must have been delirious with hunger and a wildly misplaced sense of adventure. Hallucinations, I've learned, do strange things to a man's judgment.

Lucky for us, the pressure cooker jumps in as a savior. By trapping the steam that forms early on, cookers create an "artificial atmosphere" of high pressure inside the pot. This elevated pressure raises the boiling point of the water. Inside a standard cooker, the pressure rises to about 2 atm, double the atmospheric pressure. With this pressure increase, the boiling point of water climbs up to around 121°C (250°F). Water absorbs more heat till reaching this higher boiling point, and eventually transfers it to the food, allowing it to cook faster and softens the toughest beans.

What appear to be nothing more than everyday quirks of atmospheric pressure hold lessons of lasting value which we would be unwise to ignore. At sea level, water has to withstand the entire weight of the atmosphere. It endures the pressure, absorbs the heat, resists boiling, and only transforms when it has reached its full potential. Whereas in the mountains when the pressure is lighter, water boils too soon and misses the chance to absorb all the energy it could have otherwise. Considering this analogy, what about the weight of challenges in our own lives then? Wouldn't it serve us better if we stopped seeing them as burdens and started viewing, even the heaviest trials, as instruments that push us closer to our full potential. If we can absorb the heat of hardship without boiling over in anger, frustration, or despair and if we can face the flame without cracking, only then something resolute forms within us, and only then we emerge, refined.

Even that puffed-up pack of chips, though junk as food, offers real food for thought. As we ascend the ladder of life and gain success, fame, power, or wealth, our opinion about ourselves rises too. The higher we climb, the easier it becomes to let self-importance inflate us. With rise in status, the critical voices around us grow quieter, heads get too eager to nod and spines too afraid to push back. People even start applauding our misjudgments out of fear and flattery, and we end up with our inner egos swelled like air trapped in a chip bag. If left unchecked, we risk bursting under such overinflated image of our own selves. Like a tire that rides smooth only when its pressure is just right, we too must release our air of haughtiness.

After climbing to new heights, how can one forget the plains from which he started? Instead of puffing our chests with vanity, why not choose humility? Instead of swelling with pride, why not seek modesty? Because ultimately, the greatest summit lies in staying grounded, no matter how high we climb.

Always remember: neither get humbled by suffering, nor cease to be humble in success—for it is perseverance in the lows and humility in the highs that marks a true life.

Fluid mechanics,
Many say,
It's just physics.

I see it otherwise.
To me,
It is but nature's voice.

It speaks to us
Through every bubble,
Every ripple,
Every subtle shift in air.

Listen
To a breathless climb,
A boiling pot
Or a chip bag swollen at height.

Each one shares a timeless truth,
Flowing, drifting,
Hiding in plain sight.

Strikingly bare,
Still, somehow veiled.
Such is the nature's tale.

Invisible,
Yet seen everywhere.
Indeed,
It's all in the air.

2

A Deep Dive

On a sunlit September morning in 2014, off the coast of Dahab in Egypt, a man leaned forward, exhaled, and dived into the Red Sea. When he stepped off the boat, he carried with him four years of preparation. His plan was audacious—to descend deeper than any human had ever gone on scuba.

He was not a reckless thrill-seeker though. As a former Egyptian Army officer, he had attended the U.S. Army Combat Diver course—one of the toughest programs in existence—becoming the only certified U.S. Combat Diver in the Middle East. In the years leading to this attempt, he went through scientific literature and medical research to know how the human body would react under such enormous pressure. But the trail ran cold as there were no clear answers. The only way to know was to try.

His target was staggering: nearly 330 meters (1,100 feet). At that depth the sea is not blue but black, not inviting but crushing. The pressure there exceeds 34 atm or thirty-four times greater than what we feel at the surface. This number may not do justice to what it actually feels like. To better grasp it, consider 500 pounds (225 kilograms) of weight pressing on every square inch of your body, from every direction at once.

He descended rather quickly and reached the marker at 332 meters in just 15 minutes. The ascent, however, was another story. Rising too fast meant certain death. The return took more than 13 hours—yes hours, not minutes. The scale of his achievement is hard to overstate. Divers who have ventured past 300 meters can be counted on the fingers of one hand. Many others have died trying. But why attempt such a thing at all? For Ahmad Gabr, the deepest scuba diver till date, it was the pursuit of a question: *How deep can a human being go?*

Let's take a closer look at what every diver experiences, the act of breathing underwater, and the weight that comes with it...

Water is made of H_2O molecules. It does contain oxygen but not the kind our lungs crave. To explore this aquatic realm, we have to bring our own air to breathe. With tanks of compressed air, divers can breathe just like they do on land. However, carrying air is only half the battle and the real challenge is the dreaded water pressure. On land, we don't feel the weight of the air pressing on us as it is lighter and our bodies are perfectly adapted to handle this. Things get different underwater though. Water is heavy, much heavier than air, and the deeper we go, the more weight it piles upon us. With every 10 meters (33 feet) of descent, the pressure increases by 1 atmosphere. At just 40 meters, the water is pressing on you with about 73 pounds per square inch. To put it into perspective, it is like a heavy suitcase balanced on a single square inch of skin.

Our lungs, filled with air, are compressible which makes them shrink like a balloon, when squeezed. At 10 meters below the water surface, they shrink to half their size. At a depth of 30 meters, their volume is reduced to one-fourth of the original. Breathing becomes harder, and it seems as if someone is forcefully pressing on our chest. To counteract this, diving equipment includes a device called the regulator. Connected to a tank of compressed air, it dynamically adjusts to match the surrounding pressure, for delivering air at the right pressure. Even with the regulator compensating for depth changes, our lungs still have to work really hard to draw each breath. The pressure doesn't stop at our lungs. Blood, being a liquid and thus incompressible, handles it better. But if one goes too deep, the pressure can change how oxygen travels in the bloodstream, sometimes causing dizziness.

Let's return to the question that how deep can humans safely dive? Recreational divers typically stay within 40 meters (130 feet) of depth. Beyond that, the pressure can be hazardous. Professional divers, equipped with specialized training and gear, can reach astonishing depths of up to 300 meters (1,000 feet). And Ahmad Gabr went beyond them all, into a depth no other diver has survived and returned from.

But why did it take him more than thirteen hours to return to the surface? The reason lies in how extreme pressure affects the gases inside the body. Divers breathe compressed air underwater which is a mix of nitrogen and oxygen. In deep water, the body absorbs more nitrogen than it does otherwise at the land through normal breathing. This extra nitrogen dissolves into the body's fluids and tissues such as blood, muscle, and fat. As long as the diver remains at depth, that extra nitrogen causes no harm. During the ascent however, as the surrounding pressure drops, the dissolved extra nitrogen naturally comes out of tissues and blood. If the ascent is too fast, the nitrogen escapes violently, forming bubbles inside the bloodstream and tissues. Much like the sudden fizz that bursts from a soda bottle opened too quickly. Those bubbles can block blood flow, damage nerves, and cause joint pain, paralysis, stroke, or even death. To prevent this, divers must ascend extremely slowly, stopping at specific depths—sometimes for hours—to give their body enough time to release the nitrogen safely. This slow, staged climb back to the surface is what we call *decompression*.

At such depths, the margin for error is thinner than the blade of razor one always forgets to pack. Though frankly, should the bubbles start showing up in the blood stream, I highly doubt anyone will be concerned about stubble.

How Fish Thrive Where Humans Struggle to Survive

While humans rely on specialized gear to navigate the world of water, fish seem to glide naturally, even in the crushing depths of the ocean, where the water pressure could flatten us like a pancake. Lungs, which can shrink like balloons under pressure, wouldn't work for fish. So, they got rid of them entirely and instead use gills to extract oxygen from water. Carrying no large air pockets means there is nothing for the water to compress.

Some fish have even an extra tool called a swim bladder. This gas-filled organ functions like an internal elevator. When they want to float, they add gas. When they feel like sinking, they release some. Special blood vessels around the swim bladder act like a tire pump, adjusting the gas and keeping the fish perfectly buoyant.

Let's know their secret using nothing more than a bathtub and a bottle. Place an empty, tightly capped plastic bottle into a tub full of water. The bottle pushes some of the water out of the way, that is it tries to displace it. The water, eager to reclaim its space, pushes back with a force equal to the weight of the water displaced. Since the bottle is lighter than the weight of that displaced water, the upward push wins and the bottle floats. In their usual habit of making simple things sound fancy, scientists have named this upward push *buoyancy*.

Stay in the bathtub please. The lesson isn't over yet. We're dropping a pebble next, and watch it sink, straight to the bottom. The small pebble displaces only a little water. Although it's small, the pebble is heavy for its size (denser). The upward push by that little displaced water isn't enough to counter its weight and so this time gravity wins and pebble sinks. So, the rule is if the upward force from the displaced fluid is greater than the object's weight, the object floats. If the object's weight is greater, it sinks. This is known as *Archimedes' principle* and it explains why even massive ships stay afloat. Despite their steel frames, their hollow, air-filled hulls displace so much water that the weight of the water displaced is greater than the ship's own weight.

Okay, back to the bladders—swim bladders, I mean. When a fish adds gas to its bladder, it becomes slightly lighter for the amount of water it displaces, so it floats upward. When it releases gas, it becomes heavier for its size and sinks. There is a limit for swim bladders however. They are only practical in shallow waters, typically above 500 meters (1,640 feet). At greater depths of 1,000 meters (3,300 feet) and beyond, the immense weight of the water would crush any gas inside, rendering the swim bladder useless. Worse still, if a fish rises too quickly, the expanding gas wreaks havoc on the body and the poor creature could quite literally burst (decompression).

For deep-sea fish that live far below the surface, often between 1,000 to 4,000 meters deep, the swim bladder is a death sentence of sorts. These creatures follow a different set of rulebook to withstand the extremity of depths that could easily crumple a submarine like paper in a clenched fist. Instead of gas-filled bladders, their bodies are filled with jelly-like substances and soft, flexible bones, making them nearly incompressible.

Until now, I have been guilty of casting the pressure as the bad guy of the story. Yes, it does conjure images of being overwhelmed, but what if instead, pressure isn't the villain we think it is? What if, like the best plot twists, it turns out to be the unexpected hero? Because isn't it the pressure that transforms a lump of coal into a sparkling diamond? In the same way, fluid pressure has turned once-impossible tasks into everyday conveniences so common that we hardly appreciate them. Before we begin the story of innovation, can we take a quick detour to the beach? It's worth it, I promise.

We're off to Whitehaven Beach, Australia where the sand is so soft that it feels like walking on a velvety carpet. Imagine you're wearing flip-flops. Your feet sink just a little into the sand. Now, if you switch those with a pair of sharp high heels, you will sink much deeper, right? Well, your body weight which is acting as a force hasn't changed. Force is just mass times acceleration or $F = ma$. Since neither your body mass nor the Earth's gravitational pull (acceleration due to gravity) has decided to change in the middle of the story, the force stays the same. What has changed though is the area over which that weight or force is distributed. High heels focus all your weight onto a tiny point, increasing the pressure on the sand and causing you to sink deeper. On the other hand, flat flip-flops let your weight act over a slightly larger area. As weight spreads to a bigger area, this reduces the pressure on the sand and you sink lesser.

Okay, given my track record with convincing people that hovers between dismal and nonexistent, I realize I may not have been entirely persuasive so let's try another scenario. This time you are standing in the middle of a wooden table. Of course barefoot now, because shoes on tabletop? Come on. Your entire body weight presses down on the table through the small area of your feet. If the table isn't sturdy enough and you have gained some pounds lately, you should step down quickly to avoid the disaster... or, at the very least, lie down on the table as you would on a bed. Once you lie down, your weight is now spread on the table through a much larger area—that is, through your whole body and not just your feet. This reduces the pressure on the table, and if luck holds, both the table and your bones might live to see another day.

This is the core concept of pressure. It is the force divided by the area it acts upon. Pressure = Force/Area. If the force (your weight, in this case) stays the same, reducing the area on which it acts increases the pressure, while increasing the area reduces it. The discussion of pressure leads us to one of the most ingenious insights in science known as *Pascal's law*, credited to Blaise Pascal, a 17th-century French scientist. It states,

Pressure applied to a confined fluid is transmitted equally in all directions.

If you have ever used a toothpaste tube—which, if good dental hygiene is on your radar, should be happening at least twice a day—you have already been putting this law into practice. If you don't mind, let's brush together.

So it's early morning, and you give the toothpaste tube a squeeze at the bottom. The toothpaste flows smoothly out of the nozzle at the top, just as expected. The *pressure* you *applied* at the bottom of the tube didn't just stay there. It spread throughout the toothpaste which is a *fluid confined* inside the tube. It pushed against the walls of the toothpaste tube and *transmitted equally in all directions*. Since there's only one way out i.e. the nozzle, the toothpaste had no choice but to come out from there. My kid always squeezes the toothpaste tube right in the middle. The toothpaste spreads in both directions (upward and downward) inside the tube. Only a small amount makes it out of the nozzle, while the rest retreats to the back corners. So, I usually tell him, 'Press from the bottom! That way, all the toothpaste moves upward and we can get at least a couple more brushes out of it.' But every morning, without fail, I find the tube squished in the middle again.

The same applies when we use a ketchup bottle. The kind that still has liquid inside but refuses to come out. Frustrated, we give it a squeeze, and ketchup squirts out. Sometimes onto our plates, and sometimes our shirts! Forgive me if it seems I've reduced Pascal's law to the stuff of toothpaste tubes and clumsy dinners. When in fact, it's behind much of what makes modern life, well, *modern*. From car brakes to hydraulic lifts to elevators, it does the heavy lifting. Let's see some of Pascal's muscles.

Giant Leap in Small Steps

Can you imagine lifting a car with nothing but your bare hands? Even such a thought seems impossible, doesn't it? But what if there was a machine that could take your modest effort and amplify it a thousandfold, enough to lift a truck, let alone a car? It feels fiction straight out of *Arabian Nights*, but it's real and it's called hydraulic press.

A hydraulic press consists of two cylinders—one small, one large—connected by a pipe which is filled with an incompressible fluid, usually oil. The small cylinder (or piston), is where we apply force, and the large cylinder (or piston), is where the wonder happens. When we push down on the small piston, the pressure created in the fluid is transmitted equally throughout the system according to Pascal's Law and reaches the large piston. Since the large piston has a much bigger area, the same pressure produces a far greater force. Let's make it clearer by breaking it down with numbers.

Say the small piston has an area of 2 square meters (m^2), and you push it down with a force of 10 newtons (N). As pressure is force divided by area. In this case, the pressure created in the fluid would be $10\,N \div 2\,m^2 = 5\,N/m^2$. This pressure spreads uniformly across the entire fluid. In simple terms, every part of the fluid will feel and transmit the same pressure of $5\,N/m^2$, here, there and everywhere. It would not be like that one corner gets 3, or another gets 4. So when the pressure of $5\,N/m^2$ acts on the larger piston, which has an area of, say, 100 square meters. The large piston would then be pushed up by a force of $5\,N/m^2 \times 100\,m^2 = 500\,N$. Yeah... That's right... the modest 10 newtons force on the small piston is magnified to a whopping 500 newtons force on the large piston, a 50-fold increase!

But wait—surely you know there's no such thing as a free lunch. And if you don't, the world will gladly remind you that nothing in life comes without a price. So how does hydraulic press create 500 N force out of just 10 N? While the hydraulic press might feel like magic, it doesn't create energy out of thin air. Just like everything else in the universe, it must adhere to the laws of nature and one of the fundamental laws in physics states that *energy can neither be created nor destroyed; it can only be transferred or converted from one form to another.*

The hydraulic press doesn't create energy; instead it only redistributes the one that is available. We gain a stronger output force, but at the expense of something else. The secret lies in the concept of work—a form of energy—which is simply force multiplied by distance or as mathematicians like to write Work = Force × Distance. In hydraulic press, the product of force and distance i.e. the work done, remains the same. When we push down on the small piston, we apply a small force but over a long distance. The large piston, in turn, generates a much greater force but moves only a short distance.

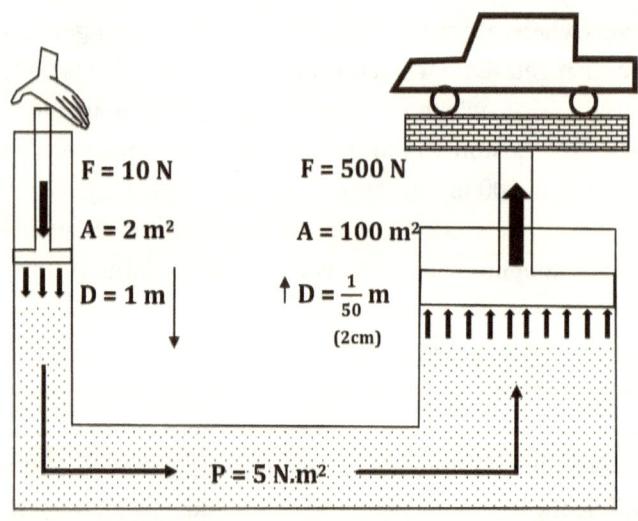

$F = 10\ N$

$A = 2\ m^2$

$D = 1\ m$

$F = 500\ N$

$A = 100\ m^2$

$D = \dfrac{1}{50}\ m$

(2cm)

$P = 5\ N.m^2$

The magic of hydraulic press

In this example, if we push down the small piston with a force of 10 N over a distance of 1 meter, the work done would be 10 N·m (10 N × 1 m). Since the total work done must stay the same, the large piston, which will produce

a force of 500 N, would only rise 0.02 meters or 2 cm
(10 N·m ÷ 500 N). This is the bargain we have to make in
which a longer, easier push on the small piston becomes a
shorter, more powerful lift on the large piston.

The beauty of hydraulic press is in its flexibility. By
adjusting the ratio of the piston areas, we can control how
much force the system amplifies. For example, if the large
piston's area were doubled to 200 square meters (m^2),
the same pressure of 5 N/m^2 would generate a force of
$5 \text{ N/m}^2 \times 200 \text{ m}^2 = 1,000$ N. Alternatively, if the large
piston's area were halved to 50 square meters, it would
generate only $5 \text{ N/m}^2 \times 50 \text{ m}^2 = 250$ N. This versatility
allows us to achieve incredible amounts of mechanical
advantage, provided the system is thoughtfully designed.
Let's shift gears, both here and on the road.

Picture yourself driving down a steep hill. As the car
picks up speed, you instinctively press the brake pedal.
The car slows down smoothly, coming to a stop exactly
when you need it. Behind the scenes, pressing the brake
pedal applies force to a tiny piston, which pushes fluid
(brake oil) inside a confined system. The pressure generated in the fluid travels instantly through the brake lines,
reaching larger pistons near the wheels. These pistons
push the brake pads against the rotors and just like that,
the car comes to a halt. Who needs Hercules when we
have Hydraulics, right?

For the same car, when we visit an auto repair shop or
a car wash, the mechanic lifts it into the air with ease using
this same principle. Also we have the hydraulic chairs in
a dentist's office or at a barbershop where a simple lever
activates the hydraulic system, allowing the chair to rise,

lower, or tilt with ease. And the story of Pascal's law continues from cranes and excavators, to tractors tilling soil, to machines unearthing minerals from the earth, and to even beyond Earth, where rocket launch systems and robotic arms repairing satellites in orbit all depend on it.

Blaise Pascal is long gone but ideas do not perish with men. His law still moves through the world. It is not some forgotten theory buried in a dusty textbook on a lonely library shelf. From toothpaste to the cosmos, it is just about everywhere. The legacy of Pascal, however, does not stop at physics. In mathematics, he laid the foundation of probability, made contributions to projective geometry, and explored infinitesimals and the summation of series (ideas that helped the development of calculus later).

Even more remarkable, and that which often gets eclipsed by his scientific genius, was Pascal's philosophical depth. Like a true erudite polymath, his intellect flowed beyond numbers. As both a philosopher and a theologian, he questioned the fragility of human existence, the limits of reason, and the eternal tension between faith and doubt in *Pensées* (French for *Thoughts*). He did not confine his work on probability to mathematical theorems or games of chance only; instead, he took a step further and applied it to the greatest question of all. One that has lingered in human minds for as long as we remember and will continue to be debated for as long as we breathe—*the existence of God.*

At the heart of his reflections lay a striking wager on belief in the face of uncertainty. His reasoning was based on weighing the potential outcomes. And when the stakes are infinite and implications endless, even the slightest chance becomes worth considering. He writes,

If you win, you win everything; if you lose, you lose nothing. Wager, then, without hesitation that He is.

On human nature, Pascal observed,

Man is but a reed, the weakest in nature, but he is a thinking reed.

No doubt we are physically weak, our bones break, hearts even more so. However, even the seemingly weak should not be considered without strength. Think of water, for example. Doesn't it yield to the slightest touch and takes on the shape of any container in which it is poured? Yet, despite all its fluidity, despite all its conformity, it doesn't shrink and remains *incompressible*. Shouldn't we strive to be like water then? Whether it is the world outside or the fights within, the struggles are real but must we collapse every time things don't go our way? No matter how heavy things get, we need to stay firm. Just like an incompressible fluid that doesn't fold.

Nevertheless, incompressibility is one thing, rigidity quite another. Rigidity is the trait of solids. It offers resistance to deformation, but its unyielding nature comes with a fatal flaw. What does not bend has to give in and so a solid pushed too far crumbles into pieces.

I find something tragically comic about those who proudly pin their inflexibilities like a medal to their chests.

To me, it is the foolishness of taking, rather mistaking, one's own obstinacy for some grand moral stance. One may admire himself as a noble rock, bravely standing midstream, only to be washed away. The wiser aim, I believe, should always be to act like a fluid—firm enough to keep your form, yet flexible enough not to shatter when the push turns into a shove.

Be incompressible like water, but just as importantly, be adaptable like the fish that navigate it. Deep-sea fish and other creatures thrive under crushing pressures by evolving to match their environment. Flexible structures and unique adaptations allow them to flourish where the rest fail. Staying attached to outdated habits or resisting change is like carrying an air pocket into the depths of ocean which is nothing but a perfect recipe for implosion. However, let there be no doubt, adaptability should never come at the cost of forsaking your principles and losing the essence of who you are. A deep-sea fish remains a fish and does not become a wild beast in the darkness of ocean.

Finally, now that you know how a hydraulic press works, remember it well that small forces get amplified but in the process something has to give in too. A gentle push can lift a car but it can't bend the laws of nature. Nature doesn't offer any shortcuts, nor does it give free rides. It only allows the chances to trade and demands the wisdom of knowing what we are trading. Doesn't the kind of exchange we see between force and distance in a hydraulic press hold even truer in our own lives?

Success, in any of its worldly forms rarely, if ever, arrives overnight with fanfare. Instead, it is built from the unglamorous grind of showing up day after day. The key is to keep pushing however small that push may seem. Every action done consistently and every effort repeated patiently eventually yields results, though not always in the short term. The piston of progress moves ever so slowly, ever so slightly, that one may swear nothing is happening at all. But it all adds up over time. Just don't be tempted by the greedy, sneaky side paths.

Though who can blame young minds, when the path laid before them is one of illusion—life hacks in place of habits, empty promises of riches without work? When all one sees is the filtered shine of social media world, it is no surprise the real unfiltered world feels like punishment. The grind feels too slow, and the discipline, too brutal. The truth is there are no soft hacks that can replace the hard hack of discipline and no shortcuts that can outpace the slow grind of patience.

Charlie Munger—whom I deeply admire and honestly, who wouldn't—is remembered as much for his wits and wisdom as for his partnership with Warren Buffett. Had he been alive, he'd have approved our "mental model" of a hydraulic press, since at one point he was asked what advice he would give to young people and this was his response:

Spend each day trying to be a little wiser than you were when you woke up. Discharge your duties faithfully and well. Step by step, you get ahead, but not necessarily in fast spurts. But you build discipline by preparing for fast spurts. Slug it out one inch at a time, day by day. At the end of the day—if you live long enough—most people get what they deserve.

3

The River Within

After walking along the beaches, climbing the mountains, and plunging to ocean depths, here we are, gazing at the mightiest river on Earth. River Amazon stretches over 4,000 miles (6,400 kilometers) and millions of gallons of water surge past every second, nourishing vast tracts of land, removing waste, and sustaining those along its bank. It is an awe-inspiring sight to say the least.

But the grandest river might not be outside... It's within. Turn the lens inward and one discovers a river just as mighty. This internal river travels approximately 12,000 miles (19,000 kilometers) daily, looping through the vessels while carrying life itself, delivering oxygen and nutrients and clearing away what no longer is required. Unlike the Amazon, which cuts through nine different countries, the impact of this river is personal, defining only one individual—*you*.

Instead of simply flowing downhill with gravity like a stream, our river is propelled by a tireless pump (the heart) that pushes it through a network of arteries, veins, and capillaries. If stretched out, this extensive structure of blood vessels would span an astounding 60,000 miles (100,000 kilometers). Yes, you read that right, it's long enough to circle the Earth more than twice. And like all great rivers, it flows according to the principles of fluid mechanics, in ways tailored to the delicate design of the human body. Let's unravel the mechanics of blood, and see how its flow mirrors the life it sustains.

You are at a bustling train station during rush hour as trains arrive and depart, shuttling thousands of passengers to their destinations. The stationmaster stands in the control room, with headphones on and eyes scanning the monitors. He directs arrivals, manages delays, and keeps everything running like clockwork. Imagine if he could never rest, not even for one second. No shifts, no coffee breaks, nothing. Just working at an endless pace, minute after minute, day after day, without any pause, and that too for years on end. That stationmaster is our heart. It works from our first breath to our last, managing the flow of blood with such precision that even the busiest train station would struggle to keep up.

The heart's job looks simple i.e. to pump blood, yet it's far from a monotonous machine. It operates as a double pump, performing two vital actions simultaneously. One side sends oxygen-rich blood through arteries, the other retrieves oxygen-depleted blood through veins and sends it back to the lungs for a fresh supply of oxygen. It could be thought of as a two-way highway—one lane delivering oxygen, the other carrying waste for disposal. Like a city's road system where smaller streets branch out to reach all neighborhoods (organs and tissues), while all traffic ultimately converges at a central hub (the heart).

Our heart beats around 100,000 times a day, pumping roughly 5 liters (1.3 gallons) of blood through the body every single minute. For perspective, take a five-liter jug full of water. Now, empty and refill that jug. Repeat the same... not once, nor twice, but 1,440 times over the course of 24 hours. That's the workload of heart. Surprisingly, the heart's impressiveness lies not just in the sheer volume of work it does, but in its adaptability. If you're lounging comfortably on the couch reading this book or sprinting to catch a train, the heart will instinctively adjust its pace to match the body's demands. When running an all-out sprint, it forces the blood to surge through arteries like a high-speed bullet train. At times of rest, it lets the blood flow unhurriedly like a local subway moving through the city in off-peak hours. During intense exercise, the heart can pump up to 30 liters (8 gallons) of blood per minute, six times its resting rate.

To keep the heart pumping, its power originates from a built-in electrical system composed of specialized cardiac cells that generate and transmit tiny electrical impulses which precisely coordinate the heart's contractions and relaxations. This system operates autonomously whether we're awake, asleep, exercising, or even unconscious.

The Transportation Network

If heart is the command center, blood vessels are the ship-ment routes that stretch to unimaginable lengths as we saw earlier. They ensure that every cell no matter where it hides, from the tips of toes to the crown of head, receives what it needs. Let's see the three key players in this system.

Arteries: The High-Pressure Highways

Arteries are the high-pressure pathways that carry blood *away* from the heart. Their walls are thick and elastic, built to withstand the strong surge of blood with every heartbeat. If you have ever seen a fire hose blasting water toward a burning building, you will know it must be both strong and flexible to handle the force without bursting. Arteries are just like that.

Besides enduring high pressure, arteries adapt by widening to increase blood flow when more oxygen is needed, while narrowing to conserve it where the demand is lower. For example, during exercise, vessels near our muscles widen, and those in less active areas, like the stomach, narrow to conserve energy. Hence why during a valiant gym session—where grown adults voluntarily grunt under metal rods in the hope of looking impressive mostly to strangers who aren't watching—our arm arteries swell, the muscles puff up handsomely, and we get that satisfying pumped look... only for the arteries to shrink back down once the body cools off.[*]

Likewise, the opposite happens after a meal, when our digestive vessels expand to deal with the mayhem

---

[*]Of course, it's not all vanity; doing exercise is good for the heart.

we have just sent down. At the same time, blood flow to the muscles is reduced, as the body—quite sensibly—prioritizes digestion. Which, perhaps, is why one should not attempt a glory jog immediately after feasting on three garlic naans and a bowl full of butter chicken. For "fitness," one can always wait until later, though that later never really shows up for many a man.

Veins: The Low-Pressure Return Roads

Once oxygen is delivered to the body, blood has to make its way back to the heart. This is where veins act as the low-pressure return roads. Veins have thinner walls than arteries as the blood they carry moves at a lower pressure. Fluids always flow from areas of higher pressure to lower pressure. Blood flow in veins (and arteries) follow the same principle relying on blood pressure, measured in millimeters of mercury (mmHg)—a concept we'll explore further later in the chapter.

Arterial pressure can reach 120 mmHg when the heart pumps, whereas venous pressure is much lower, typically around 5–10 mmHg. The right side of the heart, where blood returns, has an even lower pressure, often close to 0 mmHg. During deep breaths, the pressure in the right atrium can dip even below, creating negative pressure or vacuum. This acts like a pull or suction effect, drawing blood back to the heart to keep the cycle going. Without this pressure gradient, blood flow would stall.

The most surprising thing about veins is their ability to push blood upward, especially from the legs. They act as escalators having one-way valves, preventing blood from slipping backward. These valves make sure that even when we're standing, blood steadily climbs upward toward the heart. Without them, blood would pool in the lower body, leading to conditions like varicose veins. However, veins don't work alone. When we walk or move, the muscles in our legs act as mini pumps, squeezing the veins and helping propel blood upward. Which is why calf muscles are often referred to as the "second heart."

Capillaries: The Delivery Vans

Capillaries are the tiniest and most delicate blood vessels in our body, no wider than a strand of hair, and here is where the transaction takes place as they are positioned at the intersection of arteries, veins, and organ tissues. They receive oxygen-rich blood from tiny branches of arteries called arterioles, deliver oxygen and nutrients directly to the surrounding tissues, and then pass oxygen-depleted blood into small veins called venules, which carry it back toward the heart. We can consider them as a delivery van that drops off and picks up door-to-door packages. Their walls are so thin that substances like oxygen and glucose pass through them rather easily.

The Blood Mechanics

As we mentioned in the beginning that blood follows the principles of fluid mechanics, so now let's get to the *heart* of it. Each heartbeat generates a pressure wave; waves similar to those formed in a still pond when we drop a pebble. It is the driving force that propels blood through vessels with every beat of the heart.

In case you don't believe me—and I fully understand, I wouldn't believe myself either—please try it yourself.

Gently press your fingers against your wrist or the side of your neck. Can you feel that rhythmic thump? That's the pressure wave we are talking about and which doctors prefer to call *pulse*. If you don't feel it, it's okay, don't panic. Try the other wrist. Press a bit more gently. Drink some water. If, after all that, you are still feeling nothing, then I suggest sitting down, keep yourself very calm, and try to inform someone around that you appear to have passed away in the middle of the chapter. I commend your dedication to reading this far without your pulse.[†]

Each pulse tells how fast the heart is beating and how well the blood is moving through the body. It is effectively a personal performance tracker, offering insights about the working of heart. This is the reason in many parts of the world there are people who claim to "read" the pulse as if it were an open book. They believe its rhythm, speed, and pattern hold clues to overall health and even specific ailments.

During the flow, blood must overcome the force of resistance that opposes its movement through vessels. It can be thought of as traffic on a road. A narrow, single-lane street slows cars down and causes congestion. In contrast, a wide highway allows traffic to move smoothly and efficiently. Similarly, wider blood vessels offer less resistance, making it easier for blood to flow, whereas narrower vessels create more resistance, forcing the heart to work harder. In our bodies, gradually fatty deposits and cholesterol can build up along artery walls, narrowing the path. This is like debris piling up on a highway, forcing

---

[†]Those who have failed the pulse test should avoid attempting the next section on blood pressure.

traffic into fewer lanes and increasing the possibility of congestion. As the arteries become more restricted, the heart needs to pump harder to push blood through, raising the risk of heart disease. We saw earlier that blood vessels expand during physical activity. This actually helps keep arteries clear by improving blood flow and reducing the buildup of fatty deposits. So, by all means, keep exercising regularly, even if it's to admire yourself in the mirror now and then.

All this healthy discussion naturally leads us to one of the most important indicators of health—blood pressure. It is the force blood exerts on the walls of arteries as it flows through them. It is measured in millimeters of mercury (mmHg), but how exactly are pressure and mmHg connected? And what is even meant by millimeters of mercury?

Let's picture this simple experiment. Say you have a bowl filled with mercury, placed in open air. You take an empty glass tube—open at one end, closed at the other—and turn it upside down, putting the open end deep into the mercury. What do you expect to happen? If you think the mercury will rise up into the tube, well... it won't. It stays put because the tube isn't really *empty*. It may look like there's nothing inside, but it's already filled with air which we can't see, though it's very much there. The trapped air pushes down on the mercury, preventing it from rising.

Now if we somehow remove all the air from inside the tube and make it 'truly empty', we can create a vacuum. With no air inside the tube to push back, the surrounding atmosphere in which the bowl is placed presses down on the mercury that forces it to rise up into the empty tube.

How high can the mercury rise depends on atmospheric pressure. At sea level, air pressure is strong enough to push the mercury up to 760 millimeters. So, 1 atm is also defined as 760 mmHg.

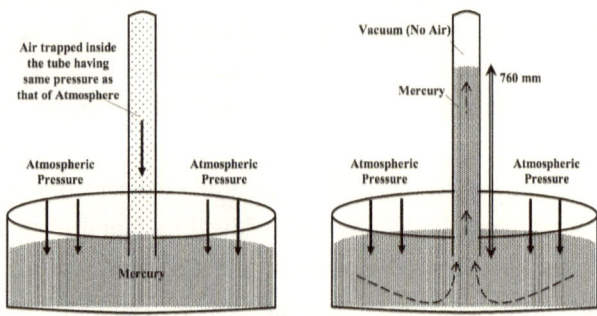

Mercury barometer

Why mercury, one might think, right? It takes us back to the early days of 17th century when scientists were figuring out a way to measure atmospheric pressure. They chose mercury for the only reason that it is incredibly dense. They could have used water and some poor souls—myself included, on a confident day—might have suggested it as well. Had they tried, they would have required a 10.3-meter (33.9-foot) glass tube, a column slightly taller than a three-story building!

By the way, it would have made quite the spectacle with learned men wobbling atop ladders and shouting things like, *Steady the base! A little higher... no, higher... HIGHER... Can't you hear me, you unpaid, underfunded PhD hopeful?* while puzzled townsfolk strolling past would've assumed a new bell tower was going up. Not ideal, really, unless they were also training for the circus.

Mercury, fortunately, saved the day. Being nearly 14 times denser than water, it brought that unwieldy column down to a more civilized 760 millimeters, without the need of acrobatics. So, the unit mmHg (millimeters of mercury) was born.

Now, back to blood pressure. The first practical blood pressure monitor, invented in 19th century, was also based on mercury (sphygmomanometer). Today modern blood pressure monitors no longer rely on mercury, the unit mmHg is still there as a tribute to this early invention. Blood pressure is measured using two numbers. Systolic pressure is the higher number, representing the force blood exerts on artery walls when the heart contracts to pump blood. A healthy systolic pressure is typically around 120 mmHg. Whereas, diastolic pressure (lower number) is the force exerted on the arteries when the heart relaxes between beats. A healthy diastolic pressure is usually around 80 mmHg. So, when the blood pressure reads 120/80 mmHg, it means the heart is generating 120 mmHg of pressure during a beat and maintaining 80 mmHg when at rest.

If someone's systolic pressure consistently rises above 140 mmHg and diastolic pressure goes over 90 mmHg, it means their arteries are getting strained. This is known as high blood pressure (hypertension). It is like forcing too much water pressure through fragile pipes. Conversely, if the systolic pressure falls below 90 mmHg and diastolic pressure drops under 60 mmHg, it means there is not enough force to keep the blood flowing effectively to vital organs. This is called as low blood pressure (hypotension).

It is as though water struggles to reach the top floors of a building, leaving one feeling lightheaded or dizzy.

Now, let's turn our attention to another crucial concept in fluid mechanics—viscosity—and see how it influences blood flow.

Viscosity is a property that indicates how thick or sticky a fluid is and how easily it flows. We may call it as the fluid's *smoothness* or *slipperiness*. Water, for example, has low viscosity as it flows easily. Sipping it through a straw is smooth. Compare that to honey which is thicker and stickier, has high viscosity, and flows more slowly. And if someone tries to sip honey through a straw, the thick consistency resists the toil, making him feel like his lungs might give up before the honey reaches the mouth. Why are you like that, *Honey!* Always making things difficult... but we love you anyway for your sweetness!

If water flows readily and honey moves sluggishly, blood behaves somewhere in between. You have probably heard the saying, *blood is thicker than water.* Figuratively, it has all the potential to start a hot debate at a family get together, in a literal sense though, it is absolutely true. Unlike water, blood is a complex fluid packed with red and white blood cells, plasma, and proteins. These components give blood its thicker texture, making it far more viscous than water. Since it is not a uniform liquid, blood viscosity can vary depending on these constituent components that may change with health, hydration levels, and the environment, which influences how easy or how hard it gets for blood to flow through our vessels.

Dehydration, for example, reduces the amount of plasma (the liquid part of blood), leaving the blood more thicker and concentrated. This increases resistance in the vessels, forcing the heart to work harder to keep blood flowing. As thicker the fluid, the harder it is to move, and the greater the pressure required. Likewise, in polycythemia vera, a rare disorder where the body produces too many red blood cells, blood becomes excessively thick, straining the heart while increasing the risk of dangerous blood clots.

On the other hand, in conditions like anemia the red blood cells drop below normal, making blood viscosity decrease. At first, this might sound appealing as thinner blood flows more easily, right? Although it does flow smoothly, on account of fewer red blood cells, the blood becomes less efficient at carrying oxygen to the tissues. The brain and other organs don't receive enough oxygen, leading to symptoms like fatigue, weakness, and even shortness of breath. So no... having fewer red blood cells isn't beneficial, it only makes it harder for the body to function properly.

Environmental factors can also have influence on blood viscosity. At higher elevations where oxygen levels are lower, the body adapts by producing more red blood cells to improve oxygen transport. Remember Chapter 1? Though crucial for survival in such conditions, it comes at a cost of blood thickening. This increases the viscosity and forces the heart to work harder. Climbers tackling Mount Everest experience this struggle firsthand. At its lofty heights, oxygen levels are barely a third of what they are at sea level. Each breath feels like sipping air through a straw. Even the simplest of the movements, like adjusting a backpack or tightening a boot, become exhausting. To compensate for the lack of oxygen, the blood starts producing more red blood cells and gets thickened, becoming harder to pump. On the outside, legs battle against the mountain's unforgiving altitude and on the inside, heart pushes against the thick blood while lungs gasp for whatever molecule of oxygen is available. In such a situation, every heartbeat becomes a battle, every breath a labor, every step an expedition, and each of them feels like scaling a miniature Everest of its own.

Blood flow, albeit a physical phenomenon, seems to reveal a greater order. It pushes through resistance, adjusts as per needs, but never stops. Isn't that, in itself, an embodiment of the wisdom to keep flowing?

Similarly, our entire body excels at adapting. Blood vessels widen or constrict, viscosity shifts with changing conditions, and the heart speeds up or slows down to meet the moment's demands. If the body instinctively knows how and when to adjust without hesitation, why do we struggle to do the same in life? Aren't we made of these very systems? If adaptation flows in our veins, why then do we resist change with such obstinate devotion in our minds?

The irony gets even harder to ignore when we see how our body carries out its role. For instance, blood flows with a purpose of delivering oxygen and nutrients. In the grander scheme of life, shouldn't we then ask ourselves what our own *oxygen and nutrients* are? What purpose fuels our flow, and where are we meant to carry it? Also consider the pressure wave that we call the pulse, whose slight changes of tempo reveal volumes about the heart's condition. Yet how seldom do we turn inward to read the pulse of our soul, where reason and virtue reside? Seek whatever binds you to nature, the greater divine order, or God—be it prayer, meditation, or even a walk among trees. Find the silence there, lest the noise of outside world cloud your inner reason.

Finally, tiny as they are, capillaries keep us alive by feeding our cells and sweeping the waste away. Likewise, beyond the body, doesn't the world depend on its own capillaries? Consider those who have carried this burden before we even understood its weight. Our parents, and especially, our mothers who are the first to rise and the last to rest. They tend to the smallest needs even before they are spoken. Their work usually goes unnoticed and uncelebrated because it is so easily taken for granted. It is such a strange nature of love. The more selfless it becomes, the tinier it gets in the eyes of those who benefit from it.

Also how about those who clean the spaces we use almost everyday—the sanitary workers who scrub public washrooms. Isn't it grueling to handle waste most of us prefer not to think about? What is more difficult: cleaning up a mess one did not make or simply offering a smile to the janitor mopping the floor? What feels harder: hauling away garbage or saying thank you to the person doing it? The least we can offer in return is a kind word and a bright smile, something that says we see them and that their work and they themselves matter. Because they most certainly do.

4

Twist in the Skies

It is some year in the 15th century. The clocks are still imprecise, maps mostly inaccurate, and ships sail with a hope that the edge of the world isn't as close as the old myths claim.

There are no satellites sending down images from space... No GPS routes... No Google Maps... No voice telling to turn left in 500 meters... No weather forecasts to warn of storms... No engines to power the ships... And even determining the exact location, that is pinpointing latitude and longitude on a featureless ocean is an arduous challenge. There is simply nothing other than the courage to dare the unknown.

For weeks, you have been sailing in such a ship, in pursuit of untold riches and lands no one can confirm exist. With no landmark in sight, you turn your gaze in all directions and find an endless stretch of blue sea, blending with the limitless blue sky at the edges. Just blue upon blue that seeps into you until you start feeling blue. The stars have been your only guides, and the winds, sole force pushing the ship forward. Without the stars, you're clueless and without the winds, powerless to move in this unmapped world of water.

Then, one day suddenly the wind vanishes completely and with that, your ship comes to a halt. The sails rest flat against the masts, as if lifeless and defeated. The only movement you could feel is the slow rocking of the waves and the only sound you could hear is the creak of wooden planks beneath your feet. The Sun beats down, baking the deck and draining the strength from your crew. Days start to blur into weeks, supplies dwindle, the water runs low, and the men grow restless with fear. You are stranded... Adrift and helpless and at the mercy of a vast void.

For sailors of the Age of Exploration—Columbus, Cook, and countless unnamed others—this wasn't just a bad dream to wake up from, instead it was their maddening reality. As they pushed beyond the familiar coasts of Europe into the vast Atlantic, they often found themselves trapped in windless zones, where the air seemed to abandon them entirely, leaving the ships stranded. In some regions, they would find steady winds that carried them westward across the ocean. To voyage home, however, they had to sail into different latitudes to find winds blowing in the opposite direction (west to east), as getting back from the same route was impossible.

Why were some regions windless, while in others the winds blew only in one direction, and then completely reverse in certain others? The sailors had no explanation.

Flow

All they knew that the air seemed to follow a set of rules which needed deciphering if they ever hoped to master the seas. Through trial and error, these explorers began to map the pathways of the winds. Over the years, they recognized a pattern. In the lower latitudes, closer to the equator, the winds flowed regularly from east to west. Farther from the equator, that is in the higher latitudes, the winds would change direction and an entirely different system blew from west to east. Whereas just above and below the equator, the winds would vanish entirely.

This discovery was monumental as it allowed ships to cross vast expanses of oceans with greater reliability. Beyond its navigational value though, these sailors had actually stumbled upon a hitherto unknown feature of our planet—a system of winds that circle the Earth in fixed directions, each bound to its place along the globe's bands of latitude. These winds still blow in the same directions, and today we call them *global wind belts*. They continue to drive weather, steer climate, and link distant corners of the world just as they did centuries ago.

But why are some regions eerily windless, while at certain latitudes the winds blow regularly only in one direction and in others, the complete opposite? Even more so, why does air move at all? Why do some days bring pleasant breeze, and others roaring winds strong enough to shake trees? How do storms form, and why do they spin? These questions pull us into the stunning world of airflow where each cloud and storm carry its own story written in pressure and temperature, and guided by the spin of the Earth. Let's follow the air...

Why Do the Winds Blow?

Wind is simply air in motion, flowing from areas of high pressure to low. Picture a balloon filled with air which is packed under high pressure. As soon as we release the balloon, the air rushes out to balance with the lower pressure outside. Wind follows the same principle, just on a far grander scale. This leads us to a question that why do different places on Earth have different air pressures in the first place?

To begin with, let's understand how air or any other fluid behaves when it is heated. As air gets warmer, its temperature increases—we all know that, right? Okay but what exactly is temperature? In simple terms, it is just a measure of how fast the molecules are moving. The faster the molecules move, the higher the temperature. So if the molecules are moving really, really fast, the temperature of that fluid (or any substance) will be extremely high. On the other hand, when we lower the temperature, we are essentially slowing the molecules down. At extremely low temperatures, their motion becomes very slow. That is why there's a limit to how much we can cool something down, because molecules can't come to a complete stop. They must stay in motion, at least a little. This limit is called absolute zero and it's −273.15°C or 0 Kelvin.*

---

*Interestingly, the upper limit to temperature is less defined. Physicists theorize a possible boundary known as the Planck temperature. It is about 1.416×10^{32} Kelvin or if you prefer unhinged numbers: 141,680,800,000,000,000,000,000,000,000,000 K. It's the point where known physics and our understanding melts. Sure, you *could* add 273 to make it Celsius—if that'll help you sleep at night.

Coming back to air, when it heats up, its temperature rises. The molecules gain kinetic energy and begin to move more rapidly. This increased motion causes the air to spread or expand and take up more space which makes the air less dense or lighter. Due to this spreading out, warm air rises from the ground and leaves an area of low pressure. Cooler air, on the other hand, stays heavier and keeps itself closer to the ground while making an area of high pressure. This rising and sinking of air, caused by difference in temperatures, creates a *pressure gradient* or simply the difference in pressure between two areas. Once there's a pressure gradient, air begins to flow from areas where cooler and heavier air has piled up (high pressure) to areas where warm air has risen and left space behind (low pressure). Apparently, I have more in common with air than I do with most people. On chilly nights, I sink deeper and deeper into my blanket, utterly convinced not to budge unless someone arrives with hot soup. But on warmer days, when the sun is out, I spring to life, bustling about with energy, though mostly looking for my misplaced socks.

Now, back to the part that why do pressure gradients exist? Our planet includes deserts, forests, mountains, oceans, and soil, each responding to Sun's heat differently. Some of these surfaces absorb heat quickly, while others take longer to warm up or cool down. To make things easier, let's keep it to two main categories—land and water. Land heats up quickly during the day because it doesn't store heat well, but as soon as the Sun sets, it loses that heat just as fast. Water, however, warms up slowly. It absorbs heat more efficiently and spreads it deeper. Once heated, it retains that warmth for a longer period of time, cooling off at a much slower pace compared to land.

Let's consider a specific example of a beach on a sunny day. The sand of beach heats up quickly under the Sun, eventually warming the air above it as well. As the warm air rises, it creates a low-pressure area. Meanwhile, the ocean, which heats up more slowly, keeps the air above it cooler and denser. This difference in pressure causes the cooler ocean air to rush in toward land, filling the gap and bringing with it a refreshing *sea breeze*. At night, the process reverses. The land loses heat quickly, cooling down much faster than the ocean. As a result, the air above the land becomes cooler and denser, sinking to create a high-pressure area. As ocean loses heat slowly and stays warmer, the air above it remains warm too and continues to rise. This time, the air flows from land to sea, forming a gentle *land breeze*.

The same phenomenon happens in valleys and mountains as well. We can picture mountain and valley as the shape of the letter V. The two slanted lines could be thought of as representing the mountain slopes, where they join at the bottom is the valley, and the upper ends of these lines represent the two mountain tops. Mountain tops and slopes heat up faster than valleys as they are fully exposed to the Sun. At higher altitudes, the atmosphere is thinner, meaning there's less air to scatter or absorb sunlight, allowing more direct heating. Also, most mountain slopes are often rocky or dry, which means they don't store heat well. They warm up quickly under the Sun but lose that heat just as fast when night falls. Valleys, on the other hand, are often shaded by surrounding peaks, which slows their warming. They have more vegetation, moisture, or soil that absorbs and holds heat, leading to more gradual temperature changes. During the day, the Sun heats the mountain slopes faster than the valley floor. The air near the slopes warms up, becomes lighter and rises, creating a low-pressure area. Cooler air from the valley rushes upward to fill the gap, forming a *valley breeze.* At night, the mountain slopes lose heat rapidly and cool down faster than the valley. The air near the slopes becomes colder and heavier, sinking into the valley creating a *mountain breeze.*

These are examples of air movement on a local scale, with winds confined to nearby areas like land and sea or valley and mountain. They are nothing but tiny ripples in the ocean of air that envelops our planet. To get the full picture, we must open our minds and look beyond local boundaries and borders to witness the spectacle that sweeps across continents and oceans... It's time we went global.

From Local to Global

It all starts with the Sun's uneven heating which gives rise to global winds and the reason for that unevenness is, once again, the Earth itself.

The spherical shape and tilt of the Earth make sunlight hit different areas at different angles during its orbit. Near the equator, the Sun's rays strike almost perpendicular (straight), concentrating their energy over a smaller, more focused area. This makes the equator much warmer. Near the poles however, the same rays arrive at a slanted angle, spreading their energy across a larger area. Since a curved surface naturally distributes incoming light more widely compared to a flat or directly facing one, so as a result the intensity of sunlight weakens and keeps the poles much cooler. To better understand this, you can hold a basketball under a flashlight. Shine the light directly at the center of the ball from some distance. The area in the middle appears bright and concentrated because the light hits head-on, but if you see toward the upper and lower edges, the light spreads out and becomes dimmer. This is exactly how the Earth's spherical shape creates uneven spread of sunlight and thus the heating. Also, at higher latitudes, sunlight must travel through more of the Earth's atmosphere before reaching the ground. This longer path forces more of the Sun's energy to be scattered and absorbed by the atmosphere, further reducing the amount of heat that reaches the polar surface.

Adding to this effect is the Earth's tilt of 23.5 degrees relative to its orbit around the Sun. At different times of the year, either the Northern or Southern Hemisphere tilts slightly toward the Sun, receiving more direct sunlight and experiencing summer, while the opposite hemisphere tilts away, bringing winter. Now, I don't expect you to still be holding the basketball—if you are, consider having a word with a professional please (or at least yourself) since sensible people would have dropped it two paragraphs ago. Though just in case you are, tilt the ball slightly, and you'll see that the intensity of the light would change on the upper and lower edges. On one edge, it increases; on the other, it decreases, depending on the side (right or left) you have tilted the ball. This tilt is what creates seasons and amplifies temperature differences across the globe.

Flow

Because of the Earth's spherical shape and its tilt, the uneven heating creates a temperature imbalance where the equator basks in intense sunlight while the poles remain cold. The cooler air from higher latitudes flows toward the equator to replace the rising warm air. One might think the air would travel in a straight line from the poles to the equator and back. Well, I have got a *twisted* tale for you.

Summer had arrived and while the rest of humanity flocked to beaches, I decided to spend it in the Arctic. An unusual choice for a holiday season, I admit, unless your idea of relaxation involves layering seventeen pieces of clothing and still getting a frostbite. But then, it was a summer like no other. Under the Midnight Sun, with weeks of continuous daylight as the sun circled the sky without ever dipping below the horizon, I walked on glaciers and immersed myself in the haunting silence of the frozen wilderness.

Now and then, I spotted polar bears in the distance, and tried saying hello but they seemed little disinterested in me. I have read somewhere they are quite proud of their fur color which, as it turns out, is transparent. Each hair shaft is hollow and colorless. It scatters and reflects visible light in such a way that makes the fur *appear* white especially against the snow. Hmm... try explaining that to a creature so devoted to the illusion of white nobility.

Anyway that was my Arctic summer and eventually it was time to say goodbye to the North pole. Standing on the icy airstrip, I pulled my coat tighter, took a deep breath of the crisp air, and climbed the stairs of the waiting plane. I settled into the seat while the captain made an announcement in a voice so mumbled it could've easily passed for engine noise: 'Ladies and gentlemen, welcome aboard. Our destination is Accra, Ghana, right on the equator. Estimated flight time: approximately 11 hours. Enjoy the flight.' I glanced at the map on the screen in front of me. The route seemed simple enough—a straight line due south—since Accra sits at roughly 0°N, 0°E, almost directly beneath the North Pole.

The flight took off smoothly and I began daydreaming about my arrival in Accra, brimming with excitement to experience the vibrant culture and the rich history of this African nation. However, as the plane descended, instead of the bustling city streets and lively markets filled with the warmth of Ghanaian hospitality, all I saw was turquoise water and palm-fringed shores. The pilot's voice crackled through the speakers,
'Welcome to Tarawa, Kiribati.'

I blinked in confusion. 'Tarawa? What Tarawa? And where's this Kiribati?' The passenger sitting beside me, casually scrolling through the phone, smirked and said, 'Kiribati is a tiny island in the Pacific Ocean about 18,000 kilometers from Accra!'

A sinking feeling hit me upon realizing how far I was from the destination. 'What's going on here?' I asked the pilot, bewildered.

'I don't know myself. I flew in a perfectly straight line from the North Pole, aiming for Accra.' he replied, clearly baffled.

'What... Oh my God!' I screamed in a fit of disbelief. 'Captain! You are not supposed to fly in a straight line!' cried another agitated passenger.

'Why not? That's how I drive the car, too. When I have to go straight, I go straight. What's wrong with it. Will someone tell me' the captain replied. His voice was more wounded now than mumbled.

Right then... here's where things went *south*. The Earth spins on its axis, counterclockwise when viewed from above the North Pole, completing one full rotation in almost 24 hours. It means the Earth turns 15 degrees every hour. Over the course of 11-hour flight, the Earth rotated 165 degrees, shifting the ground beneath the plane by nearly 18,000 kilometers at the equator, landing it in Tarawa instead of Accra. Yeah... It is a comfort to know that real-life pilots account for the Earth's rotation and constantly correct the course during flight so we don't end up sipping coconut water in the Maldives, when all we'd mentally prepared for is a soggy meat pie in Manchester.

Now, instead of an airplane, think about the air flowing in a straight line from the poles toward the equator. As the Earth spins, the ground beneath the moving air shifts. Since we're attached to the Earth's surface, rotating along with it at the same speed, we don't feel this rotating motion. To us, the ground seems absolutely still beneath our feet, even though we're constantly spinning right along with it. As a spinning observer (on a spinning Earth), when we view the movement of air from poles to equator, it appears deflected to us. Instead of moving in a straight line, the air seems to follow a curved path. In the Northern

Hemisphere, it appears to curve to the right, while in the Southern Hemisphere, it appears to curve to the left.

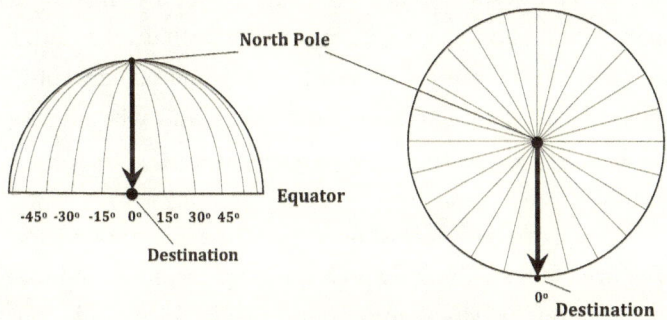

If Earth were not rotating

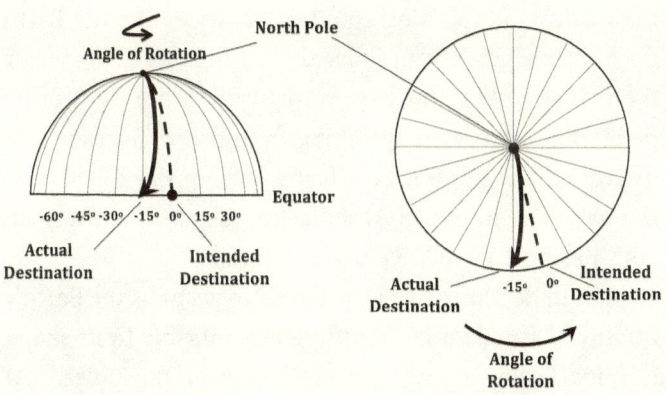

Earth rotating 15° every hour

Again, let's be clear and state this unambiguously—the air itself moves in a straight line. What makes it *appear* curved is our perspective from a rotating Earth. Since the planet is spinning, and we are spinning along with it without realizing, the moving air seems to follow a curved path from our point of view. We always like to describe things from our own frame of reference or as we perceive them, so we say that *air curves as it flows from the poles to the equator.* Whereas in reality, the air follows a straight path and it is only our rotational motion that creates this illusion. And there you have it, a fundamental concept in physics called the *Coriolis effect* that explains why moving objects appear to follow a curved path, as seen from a rotating viewpoint.

The Coriolis effect shows up at the playground too. Picture yourself standing on a spinning carousel. You toss a ball straight across to a friend on the other side, but instead of traveling in a straight line, the ball appears to curve because the carousel is spinning beneath it. Imagine the carousel is spinning counterclockwise, like the Earth does when viewed from above the North Pole. If you throw the ball straight, it will seem to curve to the right of its path. But if you were watching from below, as if looking up from the South Pole, the ball would appear to curve to the left. The direction of deflection depends entirely on your perspective relative to the rotation.

The deflection of air also appear to occur as the Earth's rotational speed varies at different points due to its shape. It spins faster at the equator than near the poles. At the equator, the surface moves at about 1,670 km/h (1,037 mph), while at the poles, the speed is effectively zero, that is the poles simply spin in place without lateral motion. As air moves across these regions, it carries the speed of the latitude it originated from but encounters ground moving at a different speed. This mismatch causes air moving toward the equator to lag behind, appearing to curve, while air moving toward the poles races ahead, curving in the opposite direction. The difference doesn't cause an immediate change but the effect gradually builds up over long distances. The simple flow of air, disrupted by Earth's rotation, turns into a complex network of global wind belts. Let's walk through these windy highways, from the calm doldrums to the icy polar easterlies.

The Doldrums (0° to 5° latitude)

Near the equator lies a zone of calm, windless air known as the doldrums. Here, the Sun's intense heat warms the air so much that it rises straight up instead of moving sideways. With no steady winds, sailors during the Age of Exploration dreaded this region, as their ships could sit motionless for weeks, under the scorching Sun, with nothing to do but wait and hope for the slightest breeze.

The Trade Winds (5° to 30° latitude)

Moving away from the doldrums, the trade winds blow steadily from east to west in both hemispheres. These winds were a lifeline for early explorers like Columbus, who relied on them to cross the Atlantic. In the Northern Hemisphere, they blow from the northeast, while in the Southern Hemisphere, they come from the southeast. Their consistency made them some of the most reliable winds on Earth, shaping centuries of maritime travel.

The Horse Latitudes (30° latitude)

Around 30 degrees north and south of the equator, we have the horse latitudes. The warm air risen from the equator cools and sinks down here, creating zones of high pressure. As air sinks, it gets diffused rather than spreading sideways to generate winds, resulting in a windless region. This zone became infamous among sailors of the past, who often found themselves stranded here. Legend has it that such sailors, desperate to lighten their ships and conserve drinking water, tossed horses overboard, thus a grim origin for the name. Today, these regions are known for their deserts like the Sahara and Atacama, which owe their arid conditions to this calm, high-pressure zone.

The Westerlies (30° to 60° latitude)

Beyond the horse latitudes, the westerlies take over. These winds blow from west to east, and are behind many of the weather patterns that affect the climate of mid-latitudes including North America, Europe, and much of Asia. They influence everything from daily forecasts to powerful storm systems.

The Polar Easterlies (60° to 90° latitude)

Near the poles, cold, dense air sinks and spreads outward, flowing from east to west. These winds, known as the polar easterlies, carry frigid air toward lower latitudes. At around 60 degrees, they collide with the westerlies, forming turbulent zones bringing blizzards, polar fronts, and intense storms in high-latitude regions.

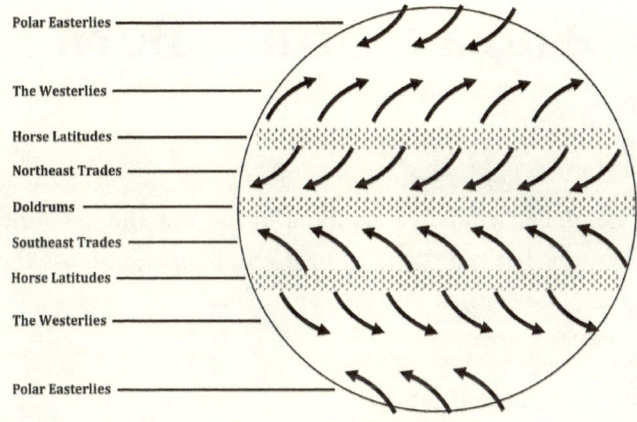

Polar Easterlies
The Westerlies
Horse Latitudes
Northeast Trades
Doldrums
Southeast Trades
Horse Latitudes
The Westerlies
Polar Easterlies

The Windy Highways

5

And a Storm is Born

The global winds we explored in the last chapter form a system that redistributes heat and moisture across the planet. Besides, it triggers one of nature's most stellar displays—storms.

The Birthplace of Storms

Our weather system revolves around two main players: low-pressure zones, where air rises, and high-pressure zones, where it sinks.

One of the most dynamic low-pressure zones on Earth lies near the equator in a narrow band roughly between 5°N and 5°S latitude. Historically, sailors called it the doldrums, an area where surface winds are weak or nonexistent. Scientists, being scientists, naturally avoided calling it something simple like "The Place with No Winds." Instead, they went with *Intertropical Convergence*

Zone (ITCZ) because heaven forbid anything in science be immediately understandable. In the ITCZ, tropical trade winds from the Northern and Southern Hemispheres converge.

Let's take a look at the wider picture as ITCZ is just one piece of a much larger puzzle. Across the broader tropical zone—stretching from 23.5°N to 23.5°S latitude—the Sun's rays hit nearly perpendicular (straight) year-round which creates intense heat. The oceans absorb this heat, warming the air above them, and this moisture-laden air eventually rises into the atmosphere. As the air rises upward, it cools down at certain heights and condenses to form clouds.

For clouds to grow larger and develop into thunderstorms, the right conditions must be in place. Clouds rely on moisture to saturate the rising air. The air must also ascend steadily as a smooth, uninterrupted rise helps the clouds develop properly. Additionally, calm winds are necessary at different altitudes; if winds at varying heights are too strong or uneven known as wind shear, they can disrupt the vertical growth of the clouds. When all these elements combine, they give rise to cumulonimbus clouds which are *the real skyscrapers*. These mighty formations can rise as high as 12–16 kilometers into the atmosphere.

All these conditions generally align in the tropics where warm tropical oceans provide abundant moisture for rising air and windless zone helps it rise steadily. While these conditions don't come together everywhere at all times, they are a routine part of tropical life particularly within the ITCZ, where heavy rainfall and storms occur nearly everyday, hence why ITCZ is called the *rain belt*.

For instance, in the Amazon Rainforest, torrential rains dominate the wet season, and cities like Manaus (Brazil) experience thunderstorms regularly. Over in West Africa, cities such as Accra (Ghana), Lagos (Nigeria), and others feel the impact of the ITCZ. In Southeast Asia, countries like Indonesia, Malaysia, and the Philippines frequently see cumulonimbus clouds, especially during the monsoon season.

The rising air sets off a domino effect. As warm air rises, it leaves behind a low-pressure zone. Nature, in search of balance, responds by drawing in cooler, denser air from the surrounding regions to fill the gap. The flow of air toward the low-pressure zone generates winds. As cooler air from higher latitudes moves toward the low-pressure zone, the Coriolis effect causes it to curve to the right in the Northern Hemisphere and to the left in the Southern Hemisphere which introduces a spiraling motion around the low-pressure center. As air continues to rush inward, the Coriolis effect deflects its path, reinforcing the spin. The inflowing air, consistently pulled from all directions and curved by Earth's rotation, eventually forms a complete rotation around the low-pressure center and the system locks into a full 360-degree spin, whirling like a giant atmospheric vortex.

Although the Coriolis effect and its associated force are fictitious from an absolute perspective, they cause storm rotation because we observe and model weather from Earth's rotating point of view. Other factors, such as the conservation of angular momentum—where air spins faster as it spirals inward, much like a figure skater pulling in their arms—further intensify the storm's rotational structure. However if aliens look at Earth from their spaceship, a tropical storm wouldn't look like a tidy vortex we Earthlings view it as. Instead, it would appear as a chaotic convergence of air masses toward a low-pressure center, with some localized swirling due to angular momentum but lacking the large-scale rotation driven by the Coriolis effect.

Okay, I know what you are thinking. So why do satellite images, taken from space, show storms as spirals? Well, because most satellites are *geostationary*. They orbit in sync with Earth's spin and stay locked above one point on the planet's surface, hence their name. They capture images from a perspective tied to the rotating frame of Earth that show the familiar spiral of thunderstorms.

Even from a non-geostationary satellite or an inertial frame of reference—*a scientific term for someone sitting in space in a fixed position and looking at us*—a storm may appear as a localized vortex due to rotational dynamics driven by angular momentum conservation. The difference however lies in how much of the observed storm rotation stems from the Coriolis effect, an artifact of Earth's rotating frame, versus these real rotational dynamics from angular momentum. In Earth-based or geostationary satellite imagery, the Coriolis effect dominates, creating a highly organized spiral structure. In a purely inertial frame, the Coriolis-driven spiral vanishes, leaving a less pronounced, less organized vortex-like appearance from angular momentum alone.

Coming back to Earth, these storms go by different names—*cyclones* in the Indian Ocean and South Pacific, *hurricanes* in the Atlantic and eastern Pacific, and *typhoons* in the western Pacific. Regardless of what we call them, they are all the same phenomenon i.e. a tropical storm sparked by solar radiation, fueled by warm oceans, and spun by Earth's rotation. Moving forward, let's just refer to them as cyclones. Simply for simplicity's sake, and not out of any regional favoritism, of course!

In the Northern Hemisphere, cyclones spin counter-clockwise, while in the Southern Hemisphere, they rotate clockwise. The trade winds usually steer cyclones west-ward toward continents. As they travel, they gain strength if the ocean beneath them remains warm, sustaining the continuous rise of air and intensifying the storm's power. In the center of a cyclone lies its most defining feature called "the eye". This is a calm, low-pressure center where surrounding air descends, like a surreal stillness in the midst of the storm's fury. The eye is encircled by the eyewall which is a ring of colossal clouds and wild winds, making it the most violent part of the storm.

The size and strength of a cyclone depend on the same key factors required for the formation of cumulonimbus clouds i.e. warm ocean waters that supply moist air for cloud development and minimal wind shear so that the storm can grow vertically without disruption. However, for a cyclone to fully develop and expand, these conditions must persist over a broader area, for an extended period, and with the added influence of the Coriolis effect to set it spinning. Cyclones typically form between 5° and 20° latitude, where the trade winds blowing from the east to the west supply the converging air that fuels the trans-formation of a low-pressure system into a rotating storm.

They don't form directly at the equator as the Coriolis force is too weak near 0° latitude to spin the storms.

When Cyclones Write History

Meteorologists categorize the power of cyclones based on their wind speeds, and let me tell you, those crazy numbers could make your head spin! A tropical storm graduates to a cyclone when its winds hit 74 miles per hour (119 kilometers per hour). To grasp how fast that really is, just stand near a highway someday as a car whips by at 120 km/h which is the maximum speed limit on highways in many countries. And yes... that's just the starting line. Cyclone clocking speeds of 96–110 mph (154–177 km/h) is a Category 2 storm, strong enough to snap trees right out of the ground. When the winds reach 130 mph (209 km/h) or higher, it becomes a Category 4 or 5 monster, capable of flattening entire towns, ripping up roads, and leaving nothing but destruction behind.

There is no shortage of sorrow in the tale of human history. Its pages are filled with dark, tragic moments. Some written by nature, others inflicted by humans upon one another. Cyclones, among nature's harshest tools, have left their share too, with indelible marks in snuffing out millions of lives and altering entire regions' destinies. These ferocious storms howl louder than jet engines with their giant waves ready to swallow everything in the path. For those trapped in its fury, it is as if the heavens are in pursuit of some long-lost vendetta, and the curtain of their doomsday has already drawn.

On an unassuming day in 1970, one such storm brewed in the Bay of Bengal. By the time it made landfall, its winds exceeded 115 mph (185 km/h), but the real devastation came from the storm surge or a huge wall of water that obliterated villages upon villages. The Bhola Cyclone, as it was named after the island it struck the hardest in present-day Bangladesh, killed an estimated 300,000 to 500,000 people, making it the deadliest tropical cyclone on record. The irony died alongside half a million souls, for in many South Asian languages, "Bhola" means innocent. A name as hauntingly misplaced as it could possibly be, almost mocking its own meaning.

Although Bhola Cyclone holds the macabre title of being the deadliest, the strongest recorded wind speeds came from Hurricane Patricia in 2015 when it reached sustained winds of 215 mph (345 km/h). Originating in eastern Pacific, it weakened before making landfall in Mexico and caused relatively minimal damage compared to its intensity, largely because it struck a sparsely populated area. Nevertheless, it still brought serious rainfall and flooding to parts of Mexico and the southwestern United States.

While maximum sustained wind speeds often come into focus, the size of a cyclone is just as important. The size is measured by the extent of cyclone's wind field, which is the distance from the center (eye) to the outer boundary (eyewall), determining the breadth of its impact. Most cyclones have wind fields spanning 300-800 km (186–500 miles) in diameter. Some storms, meanwhile, can grow to colossal proportions. Typhoon Tip, for example, which occurred in the western North Pacific Ocean in 1979. At its peak, Tip's tropical storm-force winds stretched across an astonishing 2,220 kilometers (1,380 miles). Besides being massive in size, Tip had sustained winds reaching 190 mph (305 km/h), making it a Category 5 storm. With some kindness from heavens, it weakened before hitting Japan but its sheer size caused heavy rainfall over a large area.

Calm Amidst Chaos

Everything exists in pairs. If day surrenders to night, then darkness, too, must always yield to light. The tide does not linger at its ebb forever; it surely rises again. And even the coldest winter gives way to a blooming spring. So what makes you think there wouldn't be a high for every low, be it skies or your own life? When low-pressure systems stir the chaos, high-pressure systems bring back the calm.

As we have seen, near the equator, warm, moist air rises to form giant clouds. After shedding its moisture as rain, the now dry air continues to ascend. At further higher altitudes, in the upper troposphere (about 10–15 kilometers above the Earth's surface), the air spreads out and starts drifting toward the poles. As it moves pole-ward, it gradually cools and becomes denser. By the time it reaches the subtropical regions—roughly between 23.5° and 35° latitude, both north and south of the equator—it sinks back toward the Earth, creating high-pressure zones. Descending air prevents cloud formation which results in dry and clear conditions. With little to no cloud formation, areas under high-pressure systems are home to some of the world's driest regions, marked by deserts and minimal rainfall. In the Northern Hemisphere, this belt includes the Sahara Desert in North Africa, the Middle East, and parts of Mexico and the southwestern United States. In the Southern Hemisphere, it passes through the Atacama

Desert in South America, the Kalahari Desert in Southern Africa, and regions of Australia's outback like Great Victoria Desert.

Winds and storms may be the most obvious signs of how pressure and temperature affect climates and regions but that's just the beginning as these forces cause some of the planet's most far-reaching weather phenomena. Let's look at a few of them.

Jet Streams

Jet streams are among the most fascinating features of atmospheric motion. These high-altitude *rivers of air* race around the globe at speeds reaching 400 kilometers per hour, flowing roughly 10 kilometers above the surface. Jet streams are formed due to sharp contrasts between warm and cold air, often emerging where frigid Arctic air meets much warmer air from lower latitudes. This high temperature difference generates powerful winds as the atmosphere works to balance out the energy gap. As these winds surge through the upper atmosphere, they create sweeping paths known as troughs and ridges that can slow down or accelerate weather systems. Besides affecting seasonal weather patterns, jet streams impact air travel too. Commercial flights often ride these fast-moving currents to save fuel and travel time. Conversely, flying against them results in longer journeys and higher fuel consumption.

The Monsoon

To call the monsoon the heartbeat of the region is no exaggeration, especially in the Indian subcontinent, where summer rains breathe life into the land. As the land heats faster than the surrounding ocean, a low-pressure zone forms, pulling in moisture-laden winds from the Indian Ocean. These winds bring rains as the promise of a harvest. However, monsoon is an unpredictable double-edged sword with nature sustaining life and testing the staying power in equal measure through it. Excess rain drowns crops and homes; too little leads to drought and dwindling supplies. Also what stands out about the monsoon is how it is connected with weather systems far beyond, even thousands of kilometers apart. One of its strongest influences comes from the El Niño–Southern Oscillation (ENSO), a recurring cycle of warming and cooling in the tropical Pacific.

El Niño and La Niña

Typically, trade winds blow steadily from east to west across the Pacific Ocean, pushing warm surface waters toward the western Pacific near Indonesia and Australia. This causes the water in the western Pacific to accumulate, raising sea levels and temperatures in that region. At the same time, in the eastern Pacific, cooler water rises from the depths, in a process called upwelling, that replaces the displaced surface water, keeping the region cool and nutrient-rich. This creates a temperature difference where warm waters dominate the western Pacific, while the eastern Pacific remains cooler.

During an El Niño event, the central and eastern Pacific Ocean warms abnormally. It happenes when the trade winds weaken and with less force of air holding warm water in the west, it starts flowing back toward the central and eastern Pacific. At the same time, the upwelling of cooler water in the eastern Pacific slows down or stops entirely. The consequences of this backward flow ripple across every corner of the Pacific and beyond. Regions that typically stay dry can suddenly face rains during El Niño years due to the warming of ocean waters. The west coast of South America for example, especially Peru and Ecuador, gets unusually heavy rainfall. Even parts of the western United States, particularly Southern California, may experience wetter winters, though the effects there are less consistent. In contrast, the southeastern United States, including Florida, Georgia, and the Carolinas, tends to have more reliably wet winters. This is because El Niño shifts the position of the jet stream, steering moisture-laden systems across the southern U.S. and Gulf Coast. Further away, parts of South America, like southern Brazil and northern Argentina, grapple with overflowing rivers and floods. Even East Africa including Kenya, Tanzania, and Uganda can be hit by intense rains, landslides, and flooding.

However, El Niño does the opposite in South Asia. Instead of bringing rain, it weakens the summer monsoon, leading to drier conditions and even drought. El Niño disrupts the wind pattern that normally helps pull moisture toward the Indian subcontinent (Walker circulation). When the warm waters and rising air shift eastward across the Pacific, South Asia is left under dry, sinking air, and the monsoon winds lose strength. Countries like India, Pakistan, Bangladesh, Nepal, and Sri Lanka receive less rain, affecting crops and water supplies. You must be thinking that how does then East Africa, which lies west of South Asia, end up getting *more* rain? During El Niño, the western Indian Ocean (closer to East Africa) also warms up, encouraging cloud formation and rainfall. This usually coincides with a related pattern called the Indian Ocean Dipole. When it turns positive, it means the western part of the Indian Ocean becomes warmer compared to the eastern side near Indonesia. If both sides warm equally, the dipole isn't triggered, but when the west is distinctly warmer than the eastern side, it draws more moisture toward East Africa, adding to the rainfall and increasing the risk of floods and landslides.

Next comes La Niña, the cooler counterpart of El Niño. During La Niña, trade winds get strengthened. This pushes warm surface waters even farther west and allowing more cold water to rise in the eastern Pacific. It intensifies the Indian monsoon, bringing heavier rains. El Niño and La Niña are like two ends of a climate see-saw, with the Pacific as the pivot. During an El Niño event, monsoons in South Asia weaken and cause flooding elsewhere while La Niña does the opposite by strengthening monsoons.

Why do winds blow at all? Do they merely ride their whims? Of course not. They move to seek balance, flowing from high pressure to low. Life should not be different either and must always seek balance—between the storm of ambition and the calmness of inner peace, between running after dreams and cherishing what we've already achieved, between lifting others without losing sight of our own needs. But do we notice where the winds of our life are steering us? Are we chasing a goal so fiercely that everything else is getting faded in the storm of our own making? Or are we standing still in the doldrums, motionless under the weight of inaction? Always find balance, for it is in the balance that life finds its richest meaning.

Although nature roars through the storms, it silently brings to our attention that even something so devastating begins with as innocuous elements as a patch of warm ocean water. These small imperceptible changes trigger massive forces into motion. In the same way, every choice we make, however small it may appear at the time, spins the future path of our destiny. As the famous saying goes, *The flap of a butterfly's wings in Brazil can set off a tornado in Texas.* Though not meant to be taken literally, it speaks to the same truth that small differences in the initial conditions lead to vastly different outcomes.

And yes, there is little doubt that not every "initial condition" stems from a personal choice. Some are slated way before we ever draw our first breath. Like the country we are born in, the year on the calendar, the language spoken around us, the wealth we inherit or lack thereof, the mindset with which we are raised or the absence of it, and the world's perception of our surroundings, just as much as our perception of the surrounding world. This isn't a myth but an uncomfortable truth we must acknowledge that not everyone gets a head start. Some are born on third base, while others aren't even allowed inside the stadium. A child born in the right place, in the right era, with access to the right tools, can easily see a door that others may never even know exists. Someone born in 1955 with a curious mind and early exposure to computers had a much clearer runway to become a Bill Gates than someone born the same year in a country still healing from colonization, lacking electricity, let alone a computer. There is no shame in admitting that the universe we live in is chaotic and nonlinear. We can do everything right and still falter and sometimes we stumble upon good fortune by sheer luck.

So what should one do in such a situation? Fold the sails, blame the fate, and turn to hate? Not quite. In this era of globalism and interconnectedness—noisy and chaotic notwithstanding—the gates of opportunity are no longer guarded the way they once were. Knowledge, tools, ideas, mentors, and inspiration now cross oceans in seconds. The playing field is still not level, and perhaps never will be but it is less hilly than before. We don't choose our starting point and we can't control the winds that blew at our birth, however the sails we raise and the risks we take is still our privilege. The prologue may be written by the world of randomness, but the plot twist is entirely ours to author.

And while we are on the subject of interconnectedness, let's bring into discussion the monsoon, El Niño, and La Niña. If the fall of a monsoon in India is tied to the warming of the Pacific, how can an Indian national be separate from an Australian then? Don't we all inhale the same air that travels across oceans? The same air that may have once filled the lungs of an Aboriginal ancestor centuries ago before reaching us today. The same air carrying our exhales beyond the borders we ourselves have drawn. It is humans who draw lines on the globe and call them maps; nature never draws such boundaries.

If the winds move unbothered by continents or countries, and if the sun shines without prejudice for race or creed, and if the rain falls without an eye for skin color or origin—what compels us to see divisions then? Why do we draw lines, not only on maps, but within the borders of our minds? What blinds us to the facts that nature has always known?

Perhaps the answer lies in the Coriolis effect. From the North Pole, air appears to veer to the right; from the South Pole, it seems to curve to the left. Yet all along, the air moves in a straight line and it is we who are spinning beneath it, that too, without realizing. Like the illusion of bending air, our so-called objective realities depend on our vantage point. When the biggest reality is that the differences we so tightly stick to are just perspectives, formed by where we stand or if I may say where we are born.

What we call our *truth* is nothing but our geography.

Let me elaborate this with a 'simple' example: our eating habits. The way we humans have developed them is shaped more by cultural norms, inherited traditions, and personal experiences than by any universal standard. However, we often use our own yardstick to judge others' choices without the context in which those choices are made. For instance, in some parts of the world, consumption of exotic animals is a centuries-old tradition, rooted in social practices and at times utter necessity. Historically we know that harsh climates and limited food sources left people with few options, forcing them to rely on whatever was available. Marine life in icy waters, wild game in arid deserts, or unique species in dense forests. With the passage of time, these survival-driven choices evolved into culinary traditions, becoming integral to cultural identity with a strong connection to history.

For many, however, the idea of eating exotic animals can trigger revulsion. Those who consume only certain kinds of meat, say chicken or goat, find it difficult to comprehend that why anyone would eat such creatures. Their choices are grounded in ethical or religious beliefs, leading them to see certain practices as excessive, usually without taking into account the cultural and ecological factors that make them meaningful for others.

Then, we have those who avoid any kind of meat altogether. Many vegetarians choose their diet out of compassion, health concerns, or environmental awareness. Some even refuse to consume eggs because they believe that taking a life, even a small one, is inherently wrong. They judge those who eat chicken the same way chicken-eaters judge those who eat snakes.

And the judgment does not stop there either. Even vegetarians are critiqued by advocates of broader moral perspective. 'Plants have life too,' they argue. 'Why stop at animals only? Why should plant life be excluded from the moral consideration?'

The fact is, we are all walking different paths, guided by our own moral compasses. Every food choice is more than just a matter of taste or preference. It reflects history, necessity, morality, and belief. So why do we rush to judge others when we don't know their stories? Why do we leap to conclusions about others' choices when we can never fully grasp their experiences or the world they come from?

And the example of food choice is just the surface. The issue runs much deeper than that. It runs into the very principles we have based our lives on. No doubt, it is difficult when someone questions the ideas or beliefs we hold the closest. We find comfort in the certainty of what we were taught as children, the truths handed down by our parents and their parents before them. It creates a sense of security, it feels familiar, and it forms the foundation of who we are. If someone presents an alternative perspective, it seems like an attack on ourselves and on our identity. And so we react with anger or by dismissing the other view altogether. It stops being about logic anymore. It no longer stays a rational discussion and instead becomes an existential threat to us, to everything we hold dear. And in the face of such a threat, the human instinct is never to understand, but to survive.

This is why meaningful conversations with extremists of any kind—be they race supremacists, religious zealots, or ideological purists—are nearly impossible. Like most of us, they have anchored their identity to an unchallenged ideology. When confronted with opposing views, they do not seek discourse but find refuge behind the walls of their dogmas. Rather than examining the scaffolding of their viewpoint, they try to fortify it further. To a degree that they are unwilling to even steal a glance at the terrain beyond. To a point where the mere *thought* of looking elsewhere feels like treason to their truths.

However, why blame them when we all carry our own prejudices, only pointed in different directions? Isn't it true that our most bitter conflicts arise from a simple refusal to see the world through someone else's eyes? We shut our own instead and happily settle for blindness. And all the while telling ourselves that only "we" have got the clarity. Indeed, the clarity of pitch-black darkness. And nowhere is this more entrenched than in religion and politics.

The Coriolis effect makes us realize that our paths may not be as curvy as they appear. If we take a moment to look beyond the argument, let go of the impulsive need to win by raising the volume of our voices, and instead genuinely try to comprehend what lies behind other person's views, we may find that the gap is not as wide as it seemed.

Does seeing through someone else's eyes or stepping into their shoes require surrendering our own convictions? No, it doesn't. It simply means to expand our understanding, to recognize that our perspective is just one among many, and that what we see is merely the view from where we stand in the world.

Instead of condemning others for what they eat and what they believe, or how they talk and how they walk, we would do well if we accept that what feels right to us may not be the choice of someone else. Judging others doesn't make us either wiser or more virtuous. It only builds walls where there could be bridges. It only blinds us to the rich mosaic of cultures that make the world what it is.

Be it a family quarrel or a holy war, let us not rush to decide who stands on the side of truth and who does not. And if ever our instincts compel us to declare a verdict, then it is always wiser to remember the spinning Earth beneath our feet, and that we are spinning too, though unaware. Consider where we are positioned, at the North Pole or the South Pole. Think of the possibility, no matter how small, that there *could* be more to the picture from where we are viewing it. What we uncover may not only help resolve the conflict but also open our eyes to a better acceptance of the world, its people, and our tiny place among them.

6

The Story of Flight

The moment we hear the familiar roar in the sky, our heads tilt up and eyes start scanning. We have done it since we were kids, and most of us including myself still do it almost unconsciously. However, when was the last time you looked up and wondered how does something that heavy stay up there? It's not exactly obvious. In fact, it seems a bit unfair, right?

We jump and gravity humbles us to smell the earth. Toss a rock up, and try to save your head. Yes, I know birds can fly but they have an excuse, don't they? They're light and built for it. Watching them glide and flap their wings makes flight seem natural. A plane, on the other hand, has no feathers, no flapping wings, and yet, somehow, it stays up. How? Before we get to the answer, let's rewind and go back in time. Way back to the 1700s, when people weren't flying planes but floating gently through the sky.

The year was 1783. The Montgolfier brothers stood before a buzzing crowd in France, ready to unveil their latest marvel: a giant balloon so colorful it looked like a party decoration gone wild. As they ignited the fire beneath it, the massive balloon stirred, and then began to rise slowly. It climbed higher and higher, carrying its passengers into the sky. For the first time, humans left the ground defying gravity.

What exactly though made that balloon rise? When air is heated, its molecules move faster and spread apart. This makes the air less dense meaning the same amount of space (volume) now holds less weight than it would if filled with cooler air. Inside the balloon, the fire heats the air, making it lighter than the air outside. One can imagine the balloon as a giant bubble filled with this lighter air. The cooler, denser air surrounding it pushes upward. And if the total weight of the balloon—fabric, basket, passengers, everything included—is less than the weight of the air it displaces, the balloon rises. Sound familiar? It's buoyancy, just as we saw in Chapter 2.

Humanity's first step into the skies still drift over landscapes, offering adventurous passengers breathtaking views of sunrise and sunset, and also collecting the atmospheric data. Modern hot air balloons typically fly between 1,000 and 3,000 feet (300 to 900 meters), though under special conditions, they can climb much higher. In fact, the highest recorded flight with people on board reached an astonishing 68,986 feet (21 kilometers) in 2005. However as amazing as balloons are, they can't steer, don't travel far, drift slowly, and go wherever the wind takes them. Whereas people wanted more than just floating.

There's a legend from ancient Greece, the story of Icarus that he strapped on wings made of feathers and wax, and soared into the sky. When he came too close to the sun, the wax melted, and he plunged straight into the sea. A tale that, let's be honest, needs more than just a pinch of salt... perhaps a whole bucket of it would be needed. Still, it shows just how deep our obsession with flight runs.

Centuries later, people were trying in all sorts of ways, and those early attempts were as bold as they were doomed to fail. In the 9th century, Abbas Ibn Firnas, an inventor from Spain, built a glider and leapt from a hilltop. He didn't exactly fly, but he stayed in the air just long enough to spark curiosity and draw a few gasps from onlookers. Others soon followed, with varying degrees of success. Some built gliders, while others simply strapped on wings and hoped for the best. All of them crashed spectacularly and most of them walked away with more broken bones than answers. Eilmer of Malmesbury, for example, an 11th-century English monk with a fascination for flight, made himself a pair of wings, climbed to the top of his abbey, and jumped. Perhaps he took gravity as a mere suggestion that you could politely ignore if you wore enough feathers. He crash-landed and broke both his legs. After some reflection and presumably a fair bit of bandaging, he concluded that a tail should have been added for flight stability. Now, who am I to question such an insight but had he tried again, I would've been most curious to know just how many bones he managed to save the second time around.

Anyway, we all know Leonardo da Vinci, right? We know him for painting the *Mona Lisa*, of course. But not everyone knows that he also tried to crack the code of flight.* In the 1400s, he designed flying machines of every kind ranging from bird-inspired wings to a concept not far from a modern-day helicopter. None of them ever left the ground though. By the 1700s, one thing was clear that humans couldn't fly by simply copying birds. Wings alone weren't enough and if we truly wanted to conquer the skies, we had to understand the air itself. This is where the story of aerodynamics begins. The forces that make flight possible—lift, drag, thrust—are all tied to how air behaves as a fluid. First, let's meet the brothers who turned those ancient dreams of flight into reality.

The *Wright* Way to Fly

Wilbur and Orville Wright were two brothers from Dayton, Ohio. By day they were bicycle mechanics; by night, they explored aerodynamics. It was their methodical and nearly obsessive approach to understanding flight that made them special. Unlike their enthusiastic predecessors, the Wright brothers weren't in the habit of jumping off hills and hoping that angels were on duty that day. They wanted to know why things flew or, in their case, why they *didn't*.

---

*By the way, "Da Vinci's Code" has nothing to do with it.

Armed with a knack for problem-solving, they started small by building kites and gliders. Their testing ground was Kitty Hawk, North Carolina because the place had strong winds to get them off the ground easily and also had soft sand to make their landings slightly less catastrophic.

In 1901, they tested their second major glider. Their first attempt had failed the year before, so they revised the design using the best aerodynamic knowledge available at that time. Despite those improvements, the new glider didn't perform well. It still produced far less lift than the theoretical data had promised, leaving them frustrated and disheartened. On the train ride back to Dayton, Wilbur muttered, *Nobody will fly for a thousand years.*[†] Back in Dayton, they made a bold decision to toss out the existing scientific data and conduct their own experiments. In Wilbur's words,

Having set out with absolute faith in the existing scientific data, we were driven to doubt one thing after another, until finally, after two years of experiment, we cast it all aside and decided to rely entirely upon our own investigations.

Their ambitious apparatus was a homemade wind tunnel, built right in their bicycle shop, where they tested wing shapes of several kind. This allowed them to refine their designs and better understand how air flows over wings. The Wright brothers returned to North Carolina in 1902 with a fresh and carefully engineered glider, based on their own wind tunnel data. It was still an unpowered craft, and to test it, they would launch from the sand

---

[†] Lucky for us, he was off by 998 only.

dunes of Kitty Hawk, running into the wind and letting the glider lift off as they controlled its movement. This time, their efforts paid off. The glider performed as expected, staying airborne for longer and farther than anything they had built before. They completed hundreds of successful flights and now positively believed that powered flight was also within reach.

And then came the big day. On December 17, 1903, they came back with their powered aircraft—the Flyer. To decide who would fly first, Wilbur and Orville flipped a coin. Orville won. I highly suspect Wilbur was entirely heartbroken to lose though. Anyway, Orville climbed aboard, strapped himself in and took off. And the rest, as they say, is history. Yes, yes, I know that's a cliché, but for once, could you please forgive me and actually see the moment?

The flight lasted a mere 12 seconds, covering only 120 feet (37 meters) but those 12 seconds changed our relationship with the sky forever. Skeptics (yes, they are everywhere) dismissed flying machines as impractical and dangerous but the brothers didn't deter. Over the next few years, they kept refining their design and today, the seed of *the Flyer* has grown into massive jets like the Boeing 747 and Airbus A380, aircrafts capable of carrying hundreds of passengers across oceans.

Ready to fly? Let's not rush it! Pilots also make sure to have preflight checks before takeoff, and for us it is about understanding one of the most elegant principles that fluids live by, which is:

When fluids move faster, their pressure decreases and when they move slowly, their pressure increases.

This relationship is called Bernoulli's Principle, named after Daniel Bernoulli, who first described it in the 18th century. You have every reason to question why and how this happens.

Nature follows an unshakable rule that says, *energy (in a system) cannot be created or destroyed; it only changes from one form to another.* This is known as the Law of Conservation of Energy, and it applies to everything in the Universe. We already saw it in chapter 2, in hydraulic press, right? In a moving fluid (which is our system in this case), the total energy is distributed among three main forms. First, *kinetic energy*, which comes from motion or how fast the fluid is moving. Second, *pressure energy*, which is the force the fluid applies or pushes in all directions. Lastly, *potential energy*, which depends on height and changes if the fluid moves uphill or downhill. To keep things clear, let's assume the fluid is not changing height, meaning its potential energy remains the same. This allows us to focus only on the relationship between speed and pressure. Since energy cannot be created or destroyed within a system, there is always a compromise between these forms. When a fluid speeds up, more of its energy shifts into motion, thus leaving less available for pressure. Conversely, when a fluid slows down, its motion energy decreases, allowing pressure to rise.

If this sounds a bit abstract, let's turn to a concept more familiar—money. So, imagine I've got ten dollars. Actually, don't bother imagining as it is painfully real. That is, in fact, my entire life savings. I can stuff seven bucks in my left pocket and three in my right, or six in the left and four in the right. No matter how I shuffle the money between my pockets, the total will always be ten. Same is the case with the energy (money) in a fluid. It moves between pressure (left pocket) and speed (right pocket), but the total balance (ten bucks) stays the same.

Though the caveat is that this holds true only if no outside force messes with the fluid's energy. Back to my finances, if you slip two dollars into one of my pockets, (a lovely thing to do, by the way) my total balance will go up to twelve because of your generous interference. But if you take three dollars out instead which, frankly, feels more plausible coming from you, my balance will reduce to just seven and I would be left juggling those seven bucks in my pockets. Three in the left, four in the right or one in the left, six in the right.

In ideal cases, like that of air flowing smoothly over an airplane wing, energy simply shifts between speed and pressure as Bernoulli described. However in not so ideal scenarios, external factors can affect. For example, friction between water and the walls of a pipe gradually takes away some of the energy and converts it into heat. It behaves like that friend who 'borrows' money but never returns back. This heat is then transferred away from the fluid into its surroundings, meaning the fluid loses that energy. Since this energy is no longer part of the system, it is unavailable to contribute to the fluid's speed or pressure.

In other words, the poor fluid is left juggling seven bucks just like me...

Let's try something to see for ourselves if Bernoulli was saying right or was he just making things up. Grab two sheets of paper and hold them up in front of you—one in each hand, parallel to each other—with some gap in between. Now, take a deep breath and blow air through the gap. Alright, make a prediction. Shall the papers move apart or stay where they are? Intuitively, the papers should move apart, pushed away by the gust of air you have just blown between them, isn't it?

Lo and behold! The papers will come closer to each other instead of moving apart or staying still. The air you blow between the sheets moves faster than the surrounding air. Faster-moving air creates lower pressure. Meanwhile, the still air, surrounding the papers remains at a higher pressure. The high-pressure air on the outside pushes the papers inward and pulls them together. So all it takes is two sheets of paper and a decent pair of lungs to *see* and *feel* fluid mechanics... A quick word of advice though, if someone's sitting in front of you, for humanity's sake, brush your teeth first! Otherwise, the papers will still come closer but the people around you might start moving away.

Okay, brace yourself for the next scene. This time, it's a bit terrifying one. A fire has broken out in a room. Thick and dark smoke are rising from the window. The room is so choked with smoke that seeing even a few inches ahead is impossible. A firefighter rushes with water hose in hand. He positions himself right at the edge of the window. And then... against all logic, against all sanity, instead of aiming the hose toward the burning room, he points it outward i.e. spraying water away from the fire, away from the room while sitting at the window's edge.

The crowd erupts in shock. An elderly man shouts, "Are you daft, lad? Can't you see the fire is inside, not outside? Spray the water in!" Another voice joins in from behind the fence, possibly from a man who once fought a small kitchen fire and now considers himself an expert in fire-fighting: "You're watering the walls, son! What's next? Giving us a bath?" A lady standing beside shakes her head, "This generation... With a hose in hand they think they are chief of the Fire Brigade."

Okay, let's have some clarity what the poor firefighter was trying to achieve. The powerful jet of water, when blasted outward from the mouth (opening) of window, drags the surrounding outside air along with it. This dragging of air, outside the window, increases its speed and decreases the pressure. This creates a low-pressure zone at the window. The air inside the burning room which is still at a higher pressure, gets pulled toward the window opening (lower pressure). Alongside air, smoke and heat is also pulled out of the room while the oxygen feeding the fire decreases as well. So the jet of water thrown away from the room through window opening essentially acts like a giant exhaust fan. It forces the smoke out of the choking room. Once the smoke reduces a bit, the firefighters can have better visibility which they desperately need to extinguish the fire.

There are plenty more examples around us, but let's not linger with Bernoulli for too long. We have bigger ambitions... We want to fly... And it's time to take off.

Lift: The Upward Force

Imagine you are in a car cruising down the highway. You roll down the window and stick your arm out with flat hand, palm facing the road. Does anything happen? Not much, really. In this position of hand, the oncoming air flows evenly over the top (back side) and bottom (palm side) of your hand. Since the air speed and pressure are the same on both sides (up and down), there's no significant force pushing your hand up and it stays steady.

Now, gradually tilt your hand upward, so as your palm begins to face the wind. As you do this, your hand, now at an angle, deflects the oncoming air downward. The air flowing over the top of your hand moves faster as it glides smoothly over the sloped back, while the air below moves more slowly because it meets more resistance from your palm. The air moving faster over the top creates lower pressure, while the air moving slower underneath results in higher pressure. This difference in the pressure—lower above, higher below—creates an upward force that pushes your hand. You can probably sense where this is going, right? Yes, airplanes use the same principle. Airplane wings are not made flat. They are designed with a special shape called an *airfoil*, where the top of the wing is curved, and the bottom is relatively flatter.

Let's first see what is happening above the wing. As air flows over the curved surface of the airplane wing, it gets accelerated and starts flowing faster. Fluids like air naturally move in continuous streams. When such a fluid stream encounters the curved surface of a wing, its path—called streamlines—gets squeezed closer together. In other words, the air finds itself moving in a narrower space. Now, when flowing air moves through a narrower passage, it speeds up to maintain a steady, continuous flow. So, two things are occurring: First, the curved shape of the wing forces the air into a tighter space. Second, because of this narrowing, the air speeds up.

We tackle the second part first that why does narrowing space cause air to move faster? This comes from the *continuity equation*, which states that when the available space for the flow decreases, a fluid must flow faster to keep the overall rate of movement constant. The best analogy to understand it is by considering traffic on a highway that narrows down from two lanes to one. If the same number of cars need to pass through the single lane in the same amount of time, all the cars now *must* move faster to keep traffic flowing smoothly and continuously in order to avoid a jam. Air (or any other fluid) abides by this golden traffic rule. As it is forced through a smaller space over the curved surface of the wing, it speeds up to keep the flow continuous and consistent.

Now, the first part that why doesn't the air simply shift upward into the open space above the wing, instead of "getting squeezed" in a narrow region? The answer lies in the *Coandă effect*. It is the tendency of fluids like air to stick to a curved surface rather than detaching and escaping. As the air moves over the curve of the wing, it keeps attached to it, guided along the path, instead of detaching. Of course, there's a limit, if the curvature or angle becomes too sharp, the flow can detach from the surface.

To sum it up, as air flows over the curved top of the wing, it stays attached to the streamlined surface. This attachment forces air to move through a narrower and squeezed space. This narrowing of space causes air to speed up and move faster to maintain a continuous flow. And this faster-moving air then creates a low-pressure zone above the wing.

Before we go toward the other side of the wing, let's take a small funnel and place a ping-pong ball inside it. Now blow air through the narrow end. Intuition, which has failed us earlier and will no doubt betray us again, says the ball should shoot out straight. Well, you bet. It stays inside the funnel, kind of mocking your breath. In fact, the harder you huff and puff, the more defiantly it sticks to the funnel.

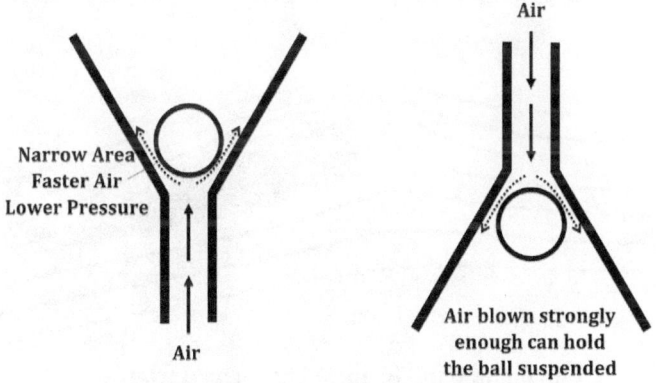

Ping-pong ball held in a funnel

It happens as the air passes from the narrow end to wider area, it finds the ball covering the space. It tries to move around the ball in what ever small space is available. As a result of moving through small space, it speeds up, and thus creates a low-pressure zone. This low pressure sucks the ball inward instead of pushing it out.

For the real party trick, you can turn the funnel upside down, pop the ball in, and place your finger on it so it doesn't fall out initially. Start blowing and then lift your finger with the flair of a stage magician (or an overconfident uncle, your choice). The ball will not drop to the floor, not if you are blowing hard enough. Due to low-pressure sorcery, it gets suspended inside the funnel. It is the perfect thing to pull out when conversation runs dry with the guests which unfortunately for me happens exactly after two and a half minutes.

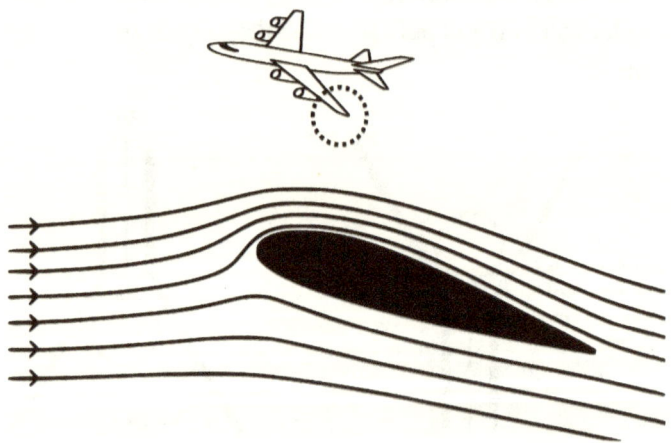

Airplane wing or *airfoil* as viewed from side

Let's go back to the bottom side of the wing where things are a lot less exciting—the grass (speed) isn't greener (faster) there. The flatter bottom surface allows air to flow at a usual pace, without the rush happening above. With no need to pick up the speed, the air moves calmly below, while creating higher pressure as compared to the top of the wing.

That's not the end of the story though. Besides being curved, the wing is also tilted slightly upward. This tilt, called the *angle of attack*, is also a contributor in generating lift. As the wing moves through the air, this angle deflects the incoming air downward, just like when you tilted your hand in the car window experiment. This deflection increases the pressure on the underside of the wing. By pushing air downward, the wing experiences an upward reaction force as well.

These two effects—curved top speeding up airflow and tilted wing deflecting air downward—combine together to create low pressure above the wing (due to faster airflow) and high pressure below the wing (due to slower airflow and air deflection). This pressure difference pushes the wing with an upward force (lift). Every curve of a wing is precisely engineered to maximize these effects so that airplanes generate enough lift to stay in the sky.

Thrust: The Forward Force

With lift keeping an airplane in the sky, next we see the force that drives it forward. Picture yourself paddling a canoe on a lake. Each time you press the paddle against the water, the canoe moves forward. This forward push is called *thrust*. It is the force that propels an object ahead. For airplanes traveling through the sea of air, jet engines or propellers act like paddles, generating thrust to push them forward through the sky.

Let's start with jet engines which work by sucking in air, squeezing or compressing it tightly, mixing it with fuel, and then setting the mixture (air and fuel) on fire. This creates a high-speed blast (jet) of hot exhaust gases that shoots out at the back of the engine. According to Newton's Third Law every action has equal but opposite reaction. The immense force of hot gases pushing out at the back (action) creates an equal force that pushes the plane forward (reaction). It is similar to when we blow up a balloon and let it go without tying the knot. The escaping air rushes out in one direction while the balloon flies in the opposite direction, though rather ungracefully, zipping about the room and knocking over teacups. A jet engine harnesses the same principle of expelling gases backward to move forward, but in a far more controlled and civilized manner. Otherwise, every flight path would resemble the wild doodles of a milk-addicted toddler.

Propeller-driven planes, on the other hand, operate a bit differently from jet engines. Instead of using exhaust gases for thrust, these planes have spinning blades, known as *propellers*, to generate forward motion. Each propeller blade is shaped like a mini airplane wing, with a rounded front and a slightly curved shape that narrows at the back. As the blades spin, they act like a swimmer's arms, pushing the air backward with each stroke. As a swimmer moves forward by pushing water behind them, the propeller's backward push on the air propels the plane forward (Yes, Newton's Third Law again). They are commonly used for smaller aircraft, where lower speeds and simpler designs are ideal.

Without enough thrust, an airplane can get in serious trouble. As it slows down, besides losing the forward push needed to keep moving, the air flowing over its wings weakens that reduces the lift too. To stay airborne, every plane needs to maintain a minimum speed, called the *stall speed.* Dropping below this speed, the wings can't produce enough lift, and the plane starts to sink.

A recent tragic example is Air India Flight AI 171, which crashed just seconds after takeoff from Ahmedabad on June 12, 2025. According to the footage and the recovered flight data recorder, the Boeing 787 rose a few hundred feet before losing thrust, and the aircraft dropped below safe speed into stall conditions before crashing.

For smaller planes, like a Cessna 172, stall speed falls between 45 and 70 mph (75 to 110 km/h). Large aircraft carriers, such as commercial jets, require much higher speeds, often between 125 and 175 mph (200 to 280 km/h), to stay above their stall threshold. The lower end of these ranges represents the minimum speed needed for steady flight, while the higher speeds are necessary at the time of takeoff to generate enough lift to overcome the plane's weight.

Drag: The Resistance

I hope you are still sitting comfortably in the car, kindly oblige me by sticking your hand out of the window again. This time, face your palm directly into the wind. You will feel a strong force pushing your hand backward, and if the car is moving really fast, holding the hand steadily becomes a real challenge, even almost impossible. This is the drag of air.

Okay, picture another situation. You are riding bicycle, happy and minding your own business, when suddenly a mischievous strong wind, probably bored and looking for entertainment, decides to blow straight into your face. Familiar, isn't it? Pedaling furiously on a windy day, hair flying, clothes fluttering, but going nowhere. Haven't we all been through this?

Drag is the force of resistance a fluid exerts on any object moving through it. The faster the object or the fluid moves, the stronger the drag becomes. For instance, the moment your car stops, the resistance on your hand disappears. Likewise, on a calm day when there is no wind, you barely notice any drag at all while cycling. Just the gentle brushing of air which is much easier to handle than when a strong wind is fighting you.

For airplanes, drag acts as a force opposing the plane's forward motion, constantly trying to slow it down. The faster a plane flies, the more drag it encounters. To counter this, airplanes are designed to be as sleek and streamlined as possible. The smooth shape reduces drag, allowing planes to fly faster and more efficiently, using less energy to maintain their speed.

There is one situation though where drag becomes essential for an airplane: landing. Unlike a car, an airplane needs to reduce its speed quickly and brakes alone can't help. This is where *spoilers*—flat panels on the wings—take effect. As the plane touches the ground, these panels flip up, disrupting the smooth airflow over the wings. By breaking up the airflow, spoilers generate a large amount of drag, helping the plane slow down rapidly. The disturbance of air flow over the wing by spoilers reduce the lift too that helps the plane stay firmly on the ground. Otherwise, the plane might bounce back up while landing.

Weight: The Pull of Gravity

Ah, and how can we forget gravity. Something that has humbled just about everything. From the airplanes to even greatest minds in quantum physics—at least for now.

One can think of the four forces of flight as directions. Lift pulls up (North), gravity pulls down (South), thrust pushes forward (East), and drag pulls backward (West). In this spectrum of contrasts, on one end we find a butterfly flapping its wings delicately, on the other, a Boeing 747 thunders like a massive metal giant weighing hundreds of tons. And despite their vastly different scales, both the butterfly flitting through a meadow and the jetliner soaring at 35,000 feet obey the same principles of balancing lift, thrust, drag, and weight to stay aloft.

Nature has spent millions of years perfecting its designs and developing solutions so elegant that we can't help but take notes. The shapes of birds, the structures of fish, and even tiny insects have helped modern aerodynamics.

When it comes to flight, birds had it mastered before we humans even imagined taking to the skies. Their wings are marvels of natural engineering, with each shaped for a specific kind of flight. Falcons, for instance, with their sleek and pointed wings, are built for speed and can dive at over 200 miles per hour to catch their prey. Albatrosses, on the other hand, have long and slender wings designed for endurance, allowing them to fly across oceans miles upon miles with barely a flap. Engineers drew inspiration from these specialized designs, such as the fighter jets like F-16 are modeled after the falcon. Their compact, responsive wings are optimized for agility and quick maneuvering during high-speed flight. In contrast, long-haul aircrafts like the Airbus A350 feature long wings that minimize drag and maximize fuel efficiency, emulating the flight of an albatross over long distances.

Also we all have probably watched a bird making tiny adjustments and angling its feathers as it flies. These imperceptible movements fine-tune airflow for stability and control. We took notice and borrowed this trick for airplanes. While an aircraft's wings lack feathers, *flaps* and *slats* serve a similar purpose by enhancing control. Flaps, as seen earlier, are located on the trailing edge of the wings and primarily increase drag, helping to slow the aircraft down during landing. Slats, positioned along the leading edge, adjust airflow over the wing to prevent stalling at low speeds, and are critical in maintaining control during takeoff and the final approach to landing.

You might have also noticed some birds announcing their presence with every flap, like pigeons whose wings create a noisy flutter. Others, like geese or swans, produce rhythmic whooshing sounds as they fly. In sharp contrast, there is a bird that when it flies, it does so without any sound. With a pair of penetrating and unblinking eyes, it hunts in near silence. Sitting motionless while scanning its surroundings with a head that swivels nearly all the way around, it appears more a ghost than a bird. In the West, it is hailed as a symbol of wisdom: solemn, silent, and scholarly. In the East however, we find its reputation quite an opposite. In parts of South Asia, calling someone by its name is like calling them an idiot. The same creature that once advised Greek goddesses in legends is now blamed for poor decisions and exam failures in Indian households. Our expert in silent flight clearly suffers from a silent identity crisis too—it's *the owl* we are taking about.

The stealth of the owl had intrigued scientists for years, until they finally uncovered its secret. Unlike most other birds, owls have serrated edges—tiny comb-like structures—along the leading edge of their wing feathers, which act like acoustic dampers. As the owl glides through the air, the serrations break up the incoming airflow into smaller streams. This reduces the chaotic eddies and whirls that typically generate noise in flight. Instead of a loud whoosh, the airflow becomes quieter, slipping over the wing with minimal disturbance. In addition, the rest of the owl's feathers are soft and velvety, which further absorb sound and prevent it from echoing off the wings. Inspired by this design, we have now developed efficient and quieter machines, reducing noise pollution and minimizing environmental disruption, including airplane engines, wind turbines, and industrial fans.

Even bees, the tiny buzzing creatures, have made us smarter. They may not seem like aerodynamic specialty, but their wing motion generates tiny vortices (swirling air patterns) that enhance lift in ways conventional flight does not. This has inspired drone design, allowing them to remain stable in unusual conditions.

Just as airplane wings borrow their shape from birds, Japan's Shinkansen bullet train takes a page from the kingfisher's book. The kingfisher dives into water with hardly a splash, owing to its pointed beak. Engineers mimicked this design, shaping the front of the trains the same way which reduced the drag and noise while improving energy efficiency.

Inspiration from nature stretches to other animals as well. Dolphins, for example, with their streamlined bodies have encouraged the design of airplane fuselages and nose cones to minimize drag. Sharks, on the other hand, have skin that is studded with tiny, tooth-like structures called dermal denticles, which channel water smoothly over its body, reducing drag as it swims through the ocean. We took this clever design for everything from swimsuits to ship hulls, creating surfaces that move faster and use less energy. In fact, nature's lessons on streamlining are now evident in modern cars like the Tesla Model S, with its grille-less design, the Mercedes-Benz EQS, shaped like a teardrop for minimal air resistance, and the Toyota Prius, built with a signature curved body to optimize airflow. It would not be wrong to say that we have only managed to scratch the surface till now. What has been explored so far is merely a drop in an ocean as vast as the Pacific.

Finding new paths demands audacity to challenge what is held as certain, and the Wright brothers were audacious without fail. When the gold standard of aerodynamics faltered in practice, they did not stick to it out of reverence and trust for tradition. They tested the winds themselves and made humanity ascend to the heavens. That is what progress is all about. It requires weighing expert opinions against reason and to cast them aside when they fall short.

It should never be construed as a slight to predecessors, when in fact, it is respect for what is true. Standing on the shoulders of giants sounds grand, and it certainly is, but only if one uses their height to spot what they missed. Lingering there and basking in the past glory without looking further is not growth. It is called camping in the huts of long-held norms or simply complacency.

It may raise some eyebrows, but if we consider the peer review system in modern research—which is supposed to sharpen ideas, catch errors, and keep science honest—then we must also admit that, more often than not, it acts like a bouncer at the doors of progress. It favors works that align neatly with prevailing theories, cite the right people, and don't ruffle too many intellectual feathers. The Wright brothers thankfully lived long enough to prove the existing knowledge and their doubters wrong. But for every such story of vindication, there are a dozen others where recognition arrived far too late. Alfred Wegener, for instance, proposed in the early 20th century that continents were not fixed but had drifted apart over millions of years. The idea was revolutionary but because he couldn't explain the mechanism, the geological community dismissed him outright. It wasn't until decades after his death that plate tectonics confirmed what he had long argued.

What's even more troubling about the current system is how, in some fields, the idea of "agreeing with peers" has devolved into a *you scratch my back, I'll scratch yours arrangement*. Favors, citations, and friendly reviews circulate within tight-knit circles. Trying to publish in a top-tier journal without a well-known name on the author list is akin to entering a private club without a sponsor.

Add onto that the ever-churning pressure of publish or perish in academia, and it's no surprise that originality or creativity has become a luxury reserved for only a privileged few. Most academics today aren't publishing to push the boundaries of science; they are publishing to survive and feed their children. They are doing it just for the sake of it, because their career, their funding, even their academic identity depends on how many papers they can produce per year. It doesn't matter if the research matters. What matters is the frequency and the volume. The result is, as expected, mediocrity—where one idea is split into three papers, where old work is reworded with just enough lipstick to make it look new, where citation counts are chased like stock prices.

There's barely any room left for slowness, for depth, for stepping back and asking real questions. All of it takes time but the system doesn't seem to have time for it anymore. And the spark that brings many brilliant minds into science slowly fades, dimmed under the bright, beaming lights of institutional pressure.

One is reminded of the case of Grigori Perelman, the reclusive Russian mathematician who, in the early 2000s, solved the legendary Poincaré Conjecture. It was one of the seven Millennium Prize Problems, a challenge that had resisted the sharpest mathematical minds for over a century. Perelman didn't submit his work to journals, nor did he present it at conferences. He quietly uploaded a series of papers directly to arXiv, an open-access preprint server, bypassing the prestige of traditional academic channels. His proof was correct. Mathematicians around the world verified it, and it was soon hailed as one of the greatest mathematical achievements of the modern era.

But when the world came calling with medals, interviews, and million-dollar prizes, Perelman wanted no part of it. He declined the Fields Medal—the highest honor in mathematics, often regarded as the Nobel Prize of the field. He also refused the Clay Millennium Prize money of one-million dollars. In his eyes, posting the proof to arXiv was enough. The work spoke for itself. And that, to him, was the only recognition that mattered. He ignored ceremonies, turned down interviews, and then one day silently walked away from mathematics altogether.

His reasons for leaving weren't rooted in eccentricity, as many first assumed, but in disillusionment. He was particularly dejected with how credit, recognition, and institutional politics were handled in the mathematical community. He openly expressed disappointment over the ethical standards, criticizing what he saw as a culture of conformity that tolerates questionable behavior.

As one commentator put it later, "He was the purest of the purists, consumed with his love for mathematics, and completely uninterested in academic politics, with its relentless jockeying for position and squabbling over credit."

In a world where even pure mathematics—a field thought to exist above human messiness—can feel like a popularity contest, what chance does a lonely idea really have? And what becomes of the minds that carry such ideas, when they discover that truth alone isn't always acknowledged?

However, almost every day, the sound of a jet draws our eyes to the skies to show us that those who dared to stand alone were often the ones who moved us all forward. Within its thunder, can we not hear the roar of the Wright brothers urging us to question the norm, calling us to dream bigger, and to set our individual path instead of following the crowd?

7

To Swing or Not to Swing

It's a cool morning of early 1800s. The lush green golf course is playing host to a group of gentlemen in tweed. They stand poised with wooden clubs in hand and brand-new rubbery ball at their feet called gutta-percha. It vaguely sounds like an exotic pudding but, in fact, is a hard material about as inspiring as a pebble.

Among them is standing a newcomer. He reaches, naively, for a fresh ball and a chuckle is heard through the tweed. "Oh please," one of them scoffs. "New balls are for regulars only. You are lucky we brought a spare today." They hand him an old ball that looked scuffed and beaten—so old that it was reeking off the dignity of those elderly people who have seen too much of life to get surprised anymore.

Shrugging his shoulders with an air of someone who's only half sure which end of the club to hold, the newbie steps up, takes a deep breath, and swings... The old, scuffed ball sails high into the air and flies far. Not just disturbingly far, but farther than any of the new balls the seasoned players had proudly sent off moments ago. A grave silence descends among the group, the kind usually reserved for funerals. "That is not possible... It shouldn't have happened," one of them mutters. The shot, off a ball that looked like it had been chewed by a horse left them in an unspoken crisis. How did it happen? Was it a fluke, a cosmic joke, or what? And then, it happened again... and then again.

Alright, alright, before you ask me for a reference or quote this tale at your next golf outing, let me offer a small disclaimer: I made it up. Wait, wait... don't get angry just yet because it isn't really far from what actually happened in the late 1800s, when golfers began noticing a strange pattern that their worn, battered balls were consistently flying farther than the fresh, shiny ones. Naturally, confusion followed and the mystery deepened that why did a rough ball outperform a pristine one straight off the shelf? Intrigued, golfers began experimenting. They deliberately scratched their new balls, smacking them against trees and rocks just to see if it made a difference. Well, much to their surprise, it really did. The scuffed balls flew farther. Something in the air, something about the way the ball interacted with it, was at work. It wasn't until the early 20th century that the riddle got solved. It was found that the bumps and the scratches on the ball were actually reducing air drag, which allowed the ball to travel farther.

These dents, now known as *dimples*, are a permanent trait of modern golf balls. But how exactly do the dimples reduce the drag? Let's clear the air.

Before diving into the details, first we see how fluids flow in general. Fluids flow in two ways: laminar and turbulent.

Laminar Flow: Smooth and Orderly

When we pour honey from a jar, it flows smoothly in layers, without swirling or disturbing its surroundings. It is an orderly procession where the fluid moves in parallel *streamlines*. No two streamlines cross each other's paths, and there is little to no mixing. We call it a laminar flow. It is calm and predictable that occurs at lower speeds, like water gently streaming through a quiet brook.

Turbulent Flow: Chaotic and Disorderly

Picture a raging river or a stormy sea, where water churns and swirls and crashes unpredictably. This is turbulent flow, a chaotic motion where the fluid moves erratically, creating *whirlpools* (circular movement) and *eddies* (small swirling currents that move against the main flow). The fluid is full of mixing and disruption instead of flowing smoothly in parallel streamlines, like when water swirls down a drain. The faster a fluid flows, the more likely it is to become turbulent.

There's one more concept we need to grasp before moving forward. Imagine we have a stationary ball with air flowing around it at some speed. When the air hits the ball, the air particles that come in direct contact with the ball surface slow down and *stick* to it. Since the ball is stationary, these air particles also come to a halt. Makes sense, right? Let's call it the first layer of air. Now, the second layer of air, just above the first, moves but much more slowly as it gets dragged by the stationary layer (the first one) beneath it. It is the same when you rub your hand across a wooden table and your skin feels friction, as the table is stationary and your hand is moving. The two adjacent layers of air also feel the similar friction. Likewise, the third layer moves slightly faster than the second, the fourth layer a bit faster than the third, and so on. Each successive layer moves more freely than the one below it. One can picture it like a concert crowd where the people closest to the stage are packed tightly and can barely move, while those farther back have more freedom to shift around. So similarly, the air right next to the ball slows down due to sticking to the surface, while air farther away continues moving at its normal speed.

This creates a thin region around the ball, where each successive layer of air varies in speed from stationary at the surface to free-flowing farther away. And this region is called the *boundary layer*, as it forms a boundary between the solid object (the ball, in this case) and the surrounding moving air. Let's break it down with numbers. Say the air is moving at 10 meters per second. The air molecules that come in direct contact with the stationary ball become almost still at 0 meters per second, just like the ball itself. The next layer of air moves slightly faster, say 1 meter per second. The third layer picks up speed to 2 meters per second, and so on, until we reach the full speed of 10 meters per second in the outermost layer. The thickness of the boundary layer is typically measured in millimeters to centimeters, depending on factors like fluid speed, size of the object, and surface roughness. More importantly, the flow inside a boundary layer could either be laminar or turbulent.

For centuries, equations of motion were known, however scientists struggled to explain why fluids near solid objects consistently refused to fit their predictions. In 1904, Ludwig Prandtl, a German physicist, came forward with this beautiful concept of the boundary layer that clarified why fluids behave differently near objects from flow farther away. It seems deceptively simple, even obvious in hindsight, and yet it eluded great minds for generations. Okay, now it's time to tee off...

Plain Golf Ball

Let's start with an undimpled or plain golf ball flying through the air. This is similar to a stationary ball placed in moving air (our earlier example), except now the ball is moving through air as well. As the ball moves forward, it collides head-on with the surrounding air. Right at the very front of the ball, the air comes to a complete stop or gets stagnant forming high pressure known as *stagnation pressure*. It is a region of high pressure because the moving air is suddenly forced to stop, and all its motion (or kinetic energy) gets converted into pressure.

From there, the air tries to move around the ball's curved surface. You can recall from Chapter 6 the *Coandă effect*, which is the tendency of fluids to stick closely to curved surfaces. Remember we said that if the curvature of the surface gets too sharp, the airflow can detach from the surface? Since the golf ball is round in shape, air finds it difficult to stick to the entire surface due to this curvature. Near the front, it clings closely, but as the flow moves toward the back, it gradually loses its grip. At some point—usually around the halfway mark—the air finally *lets go* of the surface, and the boundary layer breaks away from the ball. Once the boundary layer detaches from the ball, it leaves behind a turbulent region of swirling air called the *wake*, like the foamy chaos left behind by a speeding boat. The wake is an area of low pressure since here the air moves haphazardly or chaotically.

This contrast between the high pressure at the front (stagnation) and the low pressure behind (wake) creates a net backward force dragging the ball's forward motion, which we call pressure drag. If someone is giving you a strong push from the front, while there's hardly any push from behind, it would be difficult for you to move forward, wouldn't it?

Now, in the case of a plain golf ball, the boundary layer that forms is *laminar*, that is the air flows in smooth parallel lines along the ball's surface. The problem with a laminar boundary layer though is that it tends to separate early from the ball. Okay, are you ready for the domino effect?

The earlier the boundary layer separates from the ball, the larger the wake... The larger the wake, the lower the pressure behind the ball... The lower the pressure behind the ball, the higher the drag it faces... And need I even mention, the higher the drag it faces, the harder it is for the ball to fly farther.

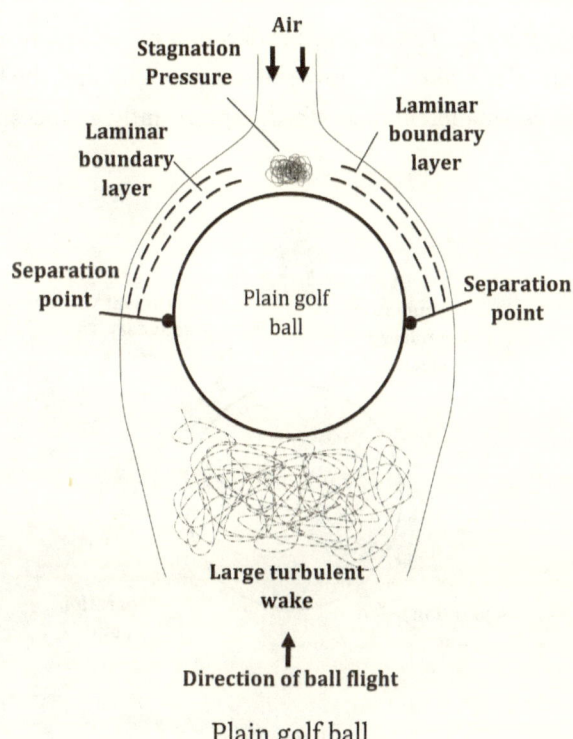

Plain golf ball

Dimpled Golf Ball

Next, we have our modern golf balls where the presence of dimples disturb the air in the boundary layer, creating tiny fluctuations and breaking up its otherwise smooth flow. As a result, the flow of air inside the boundary layer becomes turbulent, not outrageously chaotic but in a much controlled manner. The advantage of a turbulent boundary layer is that it remains attached to the ball's surface for a longer length. Instead of separating early like laminar boundary layer, it holds onto the ball's surface almost to the very back. This delayed separation of airflow results in a smaller wake. The low-pressure area behind the ball becomes smaller in size which significantly reduces the drag.

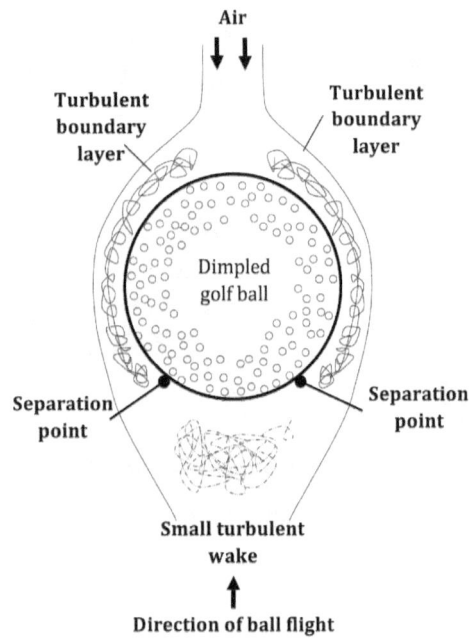

Dimpled golf ball

With a smaller wake and reduced drag, the dimpled ball can fly farther than an undimpled one, which suffers from a larger wake and therefore greater drag. The effect is so tremendous that a dimpled golf ball can travel nearly 100% farther than an undimpled ball of the same size and weight. So, if an undimpled or plain ball travels about 100-125 yards with a strong shot, a dimpled ball with the same shot could travel approximately 200-250 yards.

On a typical golf ball, we can find anywhere from 300 to 500 dimples, depending on the brand. Each dimple is roughly 3.5 millimeters long (diameter) and 0.2 millimeters deep. They are strategically arranged in hexagonal or triangular patterns, evenly covering the ball's area. Besides the depth of a dimple, the optimal gap between the two adjacent dimples is of prime importance as the air must be disturbed at exactly the right points to generate just enough turbulence in the boundary layer without going overboard and causing excessive instability.

A dimpled golf ball cuts through the heart of air with the same little resistance as a dimpled smile charms its way into the hearts. By the way, what about Tiger Woods? Doesn't he have the ultimate double advantage? One set of dimples brightening his face, and others on the golf balls, helping them fly farther. Perhaps this is his secret weapon—to have dimples everywhere!

Let's shift the scene from the golf course and head towards another green field.

December 8, 2021.
The Gabba, Brisbane, Australia.

One of cricket's greatest rivalries—The Ashes—dating back to 1882 is about to start between Australia and England. England has won the toss and opted to bat, eager to set the tone for the series with some early runs. Opening the batting for England is Rory Burns, the left-handed batter. At the top of his mark stands Mitchell Starc, Australia's left-arm pacer, who is renowned for his blistering speed and lethal swing. Starc starts his smooth run-up, reaches the crease, and bowls a quick delivery. Burns misjudges the line and exposes his leg stump. The ball swings late, darts past his bat, and crashes into the leg stump with bails flying into the air. The Gabba ignites in a storm of passion... The first ball of the Ashes and it was already a thing of beauty.

That late swing... holy smokes. That was an absolute feast to witness Starc's ability certainly, but just as much, it was an exquisite display of fluid mechanics. The way air moved around the ball made it bend like a banana leaving the batter with no chance. Rory Burns must have pondered what terrible offence had he committed against physics to deserve such targeted hostility that morning.

Let's first have a closer look at the cricket ball before seeing how it swings through the air. At the center of a cricket ball lies a solid cork which is tightly wrapped in layers of string to provide structure. Encasing this core are two halves of polished leather that give the ball its usual shine. The two leather halves are stitched together with thick threads, forming a prominent ridge along the ball's center. These raised threads are called the *seam* which divide the ball into two equal hemispheres. With six rows of stitches (three on each side), the seam acts like a miniature rudder and disturbs the airflow as the ball moves through the air. As the match progresses, another aspect of the cricket ball takes effect. At the start of the match the ball is gleaming as it is fresh out of the box. However, with every delivery it begins to lose its shine. Players try to maintain the polish of one side of the ball by rubbing it with sweat and saliva to keep it glossy while allowing the other side to roughen naturally through repeated contact with the pitch and outfield. This deliberate contrast between the polished and scuffed surfaces produces a different kind of swing. We'll come to that in a moment but first let's start with a new ball.

New Ball, Conventional Moves

Picture yourself as a bowler. You grip the ball firmly and release it with a seam so perfectly straight and upright it could serve as a moral example. You fully expect it to swing, perhaps not like Starc or Bumrah, but certainly in the general spirit of swing. But, it doesn't...

The ball travels as straight as an honest man with no movement here and there, not even a polite wobble, nothing. The batter (who is very likely me) is thoroughly unimpressed, and dispatches it to the boundary with the casual ease of someone swatting a fly. And there you stand, scratching your head, and wondering where on earth did the swing go? Probably off to find a better bowler.

Alright, don't lose heart. Let me break it to you that you are not alone. The mystery of the non-swinging ball has perplexed many a soul, man and woman alike, ranging from amateurs to professionals and the eternally hopeful in between. Though the good news is that you are reading this book and thus already on the right track, because understanding *why* it didn't swing is the first step toward making it so. And if nothing else, by the end, at least you will be able to explain a swinging yorker with the confidence of a cricket commentator, who having fallen to the very same delivery umpteen times in his career, now offers expert advice on how it should have been played. (*No, I am thinking about Rameez Raja. Why would you think that huh?*)

Okay, so you released the ball with a perfectly straight seam which looks ideal, at least on paper. Except—it isn't.

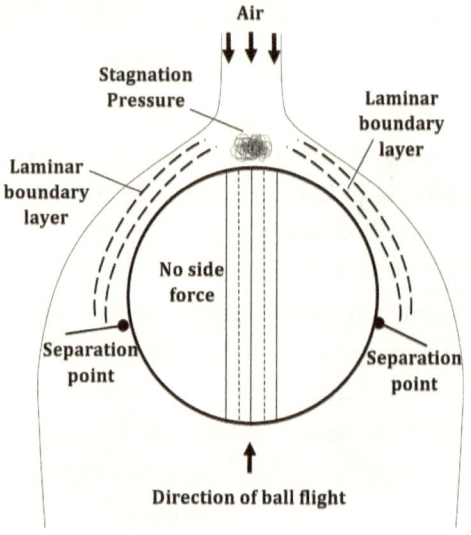

Straight seam: No side movement

A straight seam means the ball remains symmetrical as it travels through the air with three rows of stitches on one side, three on the other. Air flows evenly along both sides, tracing identical paths, leading to a symmetric airflow around it. The boundary layers detach at exactly the same point, creating equal pressure on either side. And when there is no pressure difference, there is nothing to push or pull the ball sideways and it just keeps going straight without any drama. Simply put, the ball has no reason to swing. So, why not give it one.

Swing bowling thrives on asymmetry. It happens when the ball is delivered with an angled seam, tilted slightly to one side, creating an imbalance in how the air moves around the ball. Say you position the seam at a slight angle to the left of the ball's direction of motion. Imagine an invisible line passing through the center of the tilted ball, dividing it into two halves. On the left side of this line, the raised seam is prominent. Let's call this the *seam side*. The right side of this imaginary line, which is smooth and free of the seam on the front-facing surface, becomes the *non-seam side*.

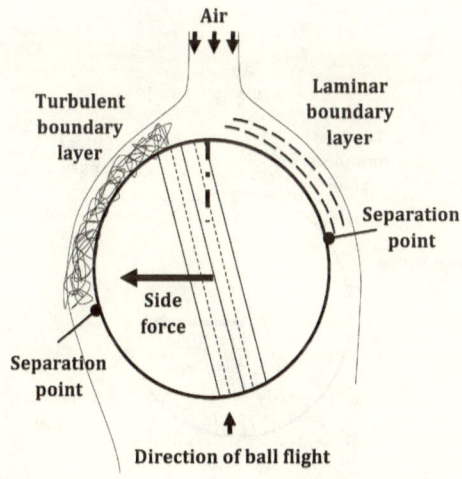

Conventional out-swinger to a right-hand batter

As the ball moves through the air, on the left side, seam (the raised thread) disrupts the smooth flow of air, causing the boundary layer to transition from laminar (smooth) to turbulent (chaotic). On the non-seam side, the surface is smooth and free of seam allowing the boundary layer to stay laminar.

We already know that the turbulent boundary layer clings to the ball's surface longer before separating while laminar layer separates earlier. This uneven separation point on both sides causes the wake to be deflected slightly toward the non-seam side. As a result, the pressure behind the ball becomes unbalanced, with the net force pushing the ball toward the seam side during flight. And that my friends is what we call *conventional swing*. It is the one produced with a shiny new ball. To right-hand batters, this left curving delivery is known as an outswinger, as it goes away from them.

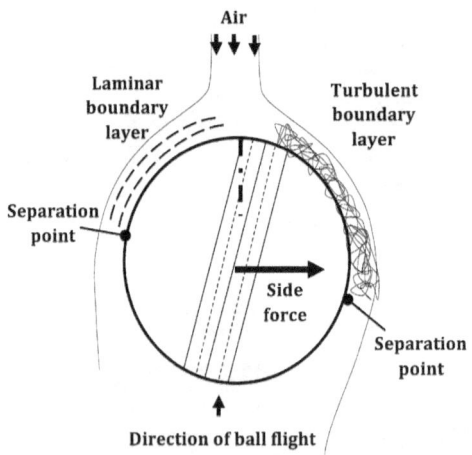

Conventional in-swinger to a right-hand batter

What if you want the ball to swing to the right? No problem, just tilt the seam to the right while releasing the ball. This small adjustment will shift everything—the seam, the turbulent boundary layer, and ultimately the swing—to the right side of the ball. To right-hander batters, this right curving delivery is known as an inswinger, as it comes into them. Tilting the seam is like steering the ball, giving you control over its direction.

Besides seam positioning, speed is also important as slow deliveries don't have enough force to generate swing; while bowling too fast creates turbulence on both sides that cancels the effect. Equally important is the ball itself. A brand-new, glossy ball with a pronounced seam is ideal for conventional swing. However, as the game wears on, the shine fades and the seam gets softened which makes the conditions for asymmetric airflow less effective.

Some bowlers are masters of outswing, others of in-swing. Only a rare few in cricketing world possessed the skill to make the new ball talk both languages with equal command. This select group includes legends like Wasim Akram, Dale Steyn, James Anderson, and the modern day great Jasprit Bumrah. Their ability to shape the ball at will, whether darting in or curving away, is what sets them apart from the rest. They belong to a different class. The elite class.

Old Ball, New Tricks

Cricket has been played since at least 1611, first recorded as a leisurely pastime in England. By 1744, it had evolved into a structured sport with written rules, and in 1877, the first-ever Test match was played. Yet, for over a century, one of the game's most potent weapons remained hidden in plain sight.

The story of *reverse swing* begins on the dusty pitches of Pakistan, where bowlers toiled with worn-out cricket balls in unforgiving conditions. Amid this struggle, they noticed something unheard-of. The ball, despite being old, was still swinging. That alone was strange on its own since such movement wasn't expected from an old ball, but even more stupefying was the fact that instead of moving toward the seam side as was usual, it was swinging in the complete opposite direction i.e. non-seam side.

At first, this bizarre movement was dismissed as a mere oddity. For a few trailblazing bowlers though, it sparked curiosity. Among them were Sarfaraz Nawaz, and later, Imran Khan, who began experimenting in search for the exact conditions and techniques needed to replicate this mysterious swing. What started as curiosity soon turned into obsession. Sarfaraz would shine one side of the ball, adjusting angles, speeds, grips and what not as if chasing a hidden pattern. Imran brought his personal blend of athleticism and analytical precision.

Gradually, the pieces began to fit. The asymmetry between the two sides, the angle of release, the seam position, the pace, all of these were ingredients for the final secret recipe. And when cooked just right, they bent the laws of swing in ways no textbook had ever explained. Yes, the reverse swing wasn't a one-time oddity. It was as real as daylight and it was a craft begging to be mastered.

By the 1990s and 2000s, Wasim Akram and Waqar Younis had elevated reverse swing to an almost mystical level. Their ability to make the ball swerve unpredictably in the dying stages of an innings turned matches on their head. It was wizardry of the highest order, enough to inspire an entire generation of bowlers. And yet, for many, the skill remained elusive. Because it was never easy. Neither then, not even now.

How does a ball that's scuffed and rough, one you'd think had lost its ability to move in the air, start swinging, that too in unexpected direction? To understand reverse swing, let's revisit the two sides of a cricket ball. At the start of the game, both sides are smooth. As the match progresses, the ball endures wear and tear, the leather starts to lose its sheen, and the seam (threads) become less pronounced. Throughout the game, fielding teams carefully polish one side of the ball, trying to maintain its surface shiny, while the other side of the ball is left to endure the scratches and constant pounding from batters. Let's call them the *shiny* and the *rough* side.

Imagine you position the seam just like in conventional out swinger, that is, slightly tilted to the left of the ball's direction of motion. Now, since the ball has two distinct sides (shiny and rough), the key is to keep the rough side on the left as well, aligning it with the seam. In other words, the rough side and the seam are on one side, while the shiny side and non-seam are on the other.

Although the shiny (non-seam) side gets maintained by the bowling team and is still smoother than the rough side, it is no longer like a new ball. Over the course of match, despite all the polish, the leather gets aged and develops slight irregularities. As air flows over this sur-face, it now creates a turbulent boundary layer instead of the smooth laminar layer when the ball was new. The turbulent layer stays attached to the ball's surface and gets separated farther down the ball.

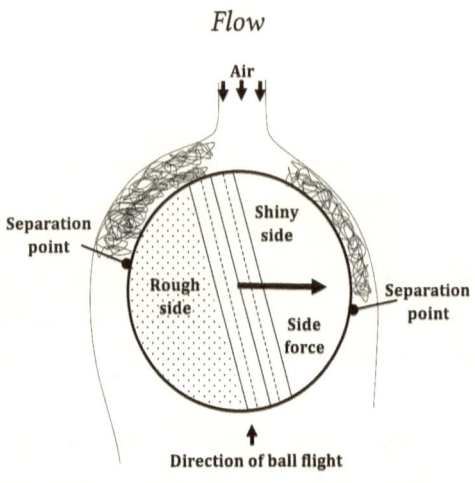

Reverse inswing

On the rough (seam) side of the ball, the story takes a chaotic turn, quite literally. The combination of the seam and the scuffed-up leather disrupts the movement of air far more aggressively. This disturbance creates erratic airflow, forming an unusually thick turbulent boundary layer on this side. Because of its excessive thickness, this turbulent layer is extremely unstable and weak. As a result it detaches from the surface much earlier than it does on the other side. This early separation on the rough (seam) side causes the wake to tilt such that the imbalance creates a net force that pushes the ball toward the other (shiny/non-seam) side. With the same seam position, the old ball now swings in the opposite direction as that of a conventional swing, and hence the name reverse swing.

Reverse swing depends heavily on speed too. At lower speeds, the conditions required for reverse swing don't develop. In these situations, the turbulent layer on the rough side doesn't become too thick to separate early, and so necessary pressure imbalance is not created. As the speed increases, typically above 85 mph (137 km/h), the airflow changes adequately and the boundary layer on the rough side becomes too turbulent to stay stable and thus separates earlier, eventually causing the pressure difference.

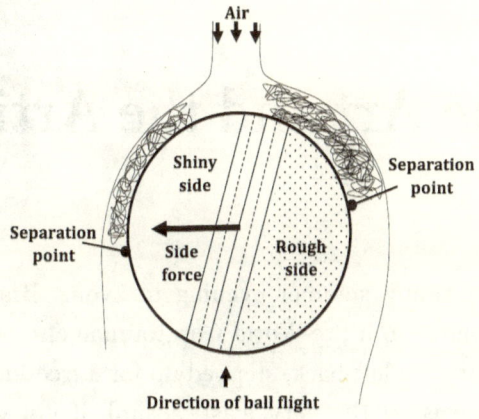

Reverse outswing

Indeed it was stumbled upon by chance, however the mastery of reverse swing is anything but a random occurrence. Let's continue playing in the next chapter...

8

The Art and the Artist

It was a balmy summer evening in Lyon. Brazil was facing France in a pre-World Cup tournament. Roberto Carlos, Brazil's left-back, stepped up for a free kick. The distance was, at the very least, absurd, if not outright impossible. Nearly 35 meters (115 feet), so far that even the most optimistic fan would hesitate to hope. A direct hit seemed utterly ludicrous to many, laughable to all. As if the distance were not enough on its own, a wall of French defenders were also standing in front, blocking the view of the goalpost.

With no clear path to the net, Roberto decided to go for it anyway. He approached the ball, knelt down, and adjusted it painstakingly, turning it several times as if looking for some particular spot on the ball. Once satisfied, he stood up and started taking steps back for his trademark long run-up.

The Art and the Artist

That's a 30-yard free kick... with a 20-yard run-up, one of the commentators quipped. He practically began running from the center line. The crowd leaned forward holding their breaths, as Roberto sprinted like a released coiled spring. He struck the ball with the full might of his legendary left leg. The ball went wide to the right of four defenders, directed toward the corner flag, and looked like a complete misfire. The defenders neither flinched, nor jumped. They simply turned their heads and watched the ball go past them. And then... out of nowhere, the ball curved—nay, it swerved—toward the goalpost. It brushed the inside of the bar and slammed into the net.

French goalkeeper Fabien Barthez, one of the finest of his era, stood frozen. His face wore an expression so unique the dictionary would have needed a new word to capture it. Something that blended amazement and the speechlessness one feels in a dream, trying to speak but unable to utter a single sound. One of the commentators said: *Barthez! Well, he was as much a spectator as I was.*

The crowd was roaring with untamed fervor and the French players could only exchange stunned glances, while Roberto was mobbed by his teammates. It was a goal for the ages. One that defied the conventional logic. One that jolted the usual senses. One that nearly bent the laws of physics with it. If soccer were an art, this free kick was its Mona Lisa, and Roberto, its Da Vinci. The only difference being it was painted in thin air. On display was the brilliance of Roberto, but behind it stood the grandeur of physics, the elegance of fluid mechanics, and in its most sublime detail, the stellar beauty of *Magnus effect*. Let's chase it...

Say, we have a round soccer ball traveling through the air. As we saw in Chapter 7, when air flows symmetrically around a ball, the forces on either side balance out, keeping it on a straight path. A straight path is indeed useful, but not exactly exciting. So how about we make it a bit thrilling by adding a little spin to the ball? If we spin the ball along the vertical direction (y-axis), counterclockwise as viewed from above, it will throw the symmetry out the window and ball's interaction with the air will change completely on either side.

As the ball spins counterclockwise, its surface drags the incoming air along with it. On the left side, the spin works with the airflow. The ball's surface moves in the same direction as the air. It acts like a conveyor belt in airports that speeds up passengers who are already walking in that direction. This makes the air on this side move faster, as spin and airflow combine forces. On the right side, the ball's surface moves against the airflow. Picture some pretending-to-be-cool teen trying to climb up a downward-moving escalator in a mall. Does he reach the top or get stuck awkwardly halfway, trying not to look like he's losing? In the same way, this opposing motion slows the air down on the right side of the ball, creating a clear difference in airspeed between the two sides.

I hope you can sense where we are heading, right? The difference in airspeeds leads to a pressure imbalance between the two sides of the ball. Faster-moving air on the left side creates lower pressure, while slower-moving air on the right side results in higher pressure. This imbalance in pressure pushes the ball from the higher-pressure to the lower-pressure side, causing the ball to curve and veering away from a straight path. In the language of fluids, we call it the *Magnus effect*. Here, since the ball is spinning counterclockwise, it curves to the left as higher pressure on the right pushes it in that direction. If the spin were reversed (clockwise), the ball would curve to the right, following the same principles but in the opposite direction. The faster the spin, the greater the pressure difference, and the more exciting the curve of the ball.

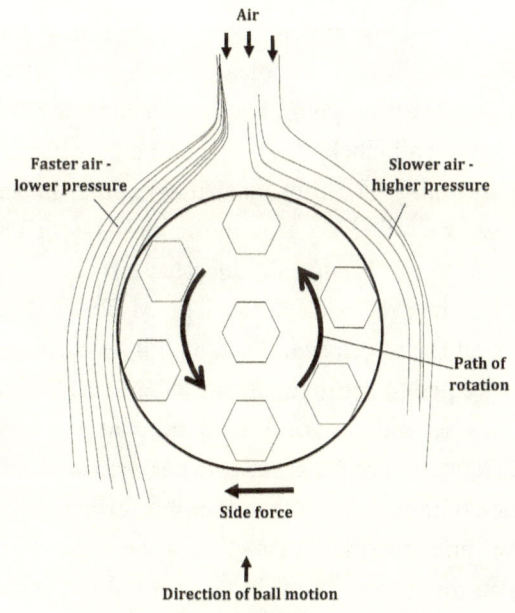

The Magnus effect

Let's revisit the free kick. With the outside of his left foot, Roberto delivered a robust strike that imparted a counterclockwise spin (when viewed from above) to the ball. The moment it left the ground, the ball started rotating rapidly along the vertical axis. This rotation disrupted the airflow around it, creating a pressure difference that pushed the ball toward the lower-pressure side and caused it to change its path mid-air. The ball bent toward the goal tracing a trajectory no one could have anticipated. If theorists define the laws, then in that moment, Roberto was the experimentalist, who brought the theory to life.

How about we take the idea of Magnus and put it to test beyond the soccer field? Let's go to Dubai once again. Uh... don't worry. I have a new plan this time. We are skipping the beach and going somewhere else.

Before we go, can you please check what time it is? No, don't check it on your smartwatch or phone. Look at an analogue wall clock. Do you see the thin hand counting the seconds? The thin hand that counts the seconds completes one full circle around the dial—say, from 12 to 12—in 60 seconds. We usually measure circular motion in *degrees*, where a full circle is 360 degrees. There is another unit as well known as *radians*. One full revolution of 360 degrees is equal to approximately 6.28 radians. This means the clock's second-counting hand spins at 6 degrees per second (360 ÷ 60) or 0.105 radians per second (6.28 ÷ 60). Let's stick to radians per second for our further discussion.

Now, imagine that instead of a slow-moving clock hand, you move your own finger in a full circle over the clock face, from 12 back to 12, in just one second. That means you are spinning your hand at 6.28 radians per second. If you can go faster and complete two full circles in one second, your spin rate doubles to 12.56 radians per second. Buckle up, we are taking this concept of spinning to new heights.

Here we are, standing atop Burj Khalifa in Dubai, the tallest skyscraper in the world. At 828 meters, it towers over everything, piercing the sky like a needle. You have a soccer ball in your hands and you drop it from that height. The ball falls straight down and lands near the foot of the Burj Khalifa, exactly where you'd expect.

Now it's my turn. Being the cool dude I am (self-proclaimed), I go for the same but with a spin. Before releasing my soccer ball, I grip it with both hands and twist strongly to generate a spin, say 20 radians per second, a speed that anyone with a decent grip can easily achieve. If you want to see this spin, you can rotate your finger on a clock dial almost three circles every second. As the ball plummets, instead of simply falling straight down, because of the spin it will move sideways as well, pushed continuously by the Magnus force. Let's assume the ideal conditions that there is no wind and air drag is minimal. Take a wild guess, where do you think the ball would land? As a hint, it is definitely not near your soccer ball. As the ball drops from the lofty height of 828 meters, the Magnus force keeps nudging it, pushing it farther and farther away from the tower. By the time it finally hits the ground, it's not just a few meters off, neither a kilometer or two. It will land more than 10 kilometers away. Yeah... that's right. Recover your spinning head with a raised eyebrow and a cup of tea, meanwhile I'll go fetch the poor ball.

Alright, I'm back. No luck finding the ball, it is probably lost to the sands of time (and the actual sand of desert). So I grabbed a new one on my way back. Of course I paid for it, only begrudgingly though. I have done many questionable things in life, but retail crime isn't one of them. Though judging by the look on your face... okay, never mind.

Let's have a closer look at the ball. For most of the 20th century, soccer balls followed a traditional design of 32 leather panels (20 hexagons and 12 pentagons) stitched into a near-perfect round shape. Reliable as it was, this classic version had a flaw. In rain, the leather soaked up water, turning the ball heavy and sluggish. By the early 2000s, as technology advanced, manufacturers like Adidas and Nike began to rethink the ball's design. They reduced the number of panels and replaced stitching with thermal bonding (a technique that fuses the panels seamlessly).

No soccer ball, no matter how advanced, is perfect though and has tiny ridges, dimples, or micro-bumps. These imperfections dictate the flow of air around the ball which affects drag and stability. Such textured surfaces, paired with fewer seams, oftentimes hold an element of surprise for everyone. Let's find it out...

It was January 30, 2008. Old Trafford, the fortress of Manchester United, was buzzing with energy as the Red Devils were facing Portsmouth in a Premier League clash. Cristiano Ronaldo, already a rising superstar, was standing over the ball just outside the penalty area to take a free kick. The crowd had seen him score from similar positions before, but this one looked tough as the angle was awkward. Wayne Rooney stepped up beside Ronaldo, leaning in to say something. A brief exchange took place between them. No one knew exactly what was said but whatever it was, it only sharpened Ronaldo's focus and widened his eyes even further.

He took his hallmark stance—legs apart, chest puffed, eyes focused—a posture that radiated authority. And struck the ball with his right foot, just below its center, launching it through the air with smashing speed. The ball appeared to sail harmlessly over the crossbar, drawing groans of disappointment. But then, all of a sudden, the ball bent sharply to the left and dipped as if an eagle were diving on its prey. Portsmouth goalkeeper David James barely reacted. The ball was past him in a blink, ripping into the top corner of the net.

Old Trafford skipped half a heartbeat. Then the stands, and every heart within, trembled with wild exhilaration. What made this goal unforgettable was the path the ball took. It curved and it dipped and it zigzagged, as if it were a drunken man on a slippery dance floor.

To make sense of the madness, let's focus on the ball. For a spinning ball, airflow behaves in a predictable way as we've seen earlier. But when the ball has little to no spin, things get rather undetermined. Its movement is then influenced by the ball's seams and textured surface (those tiny ridges and grooves). These features disturb the airflow unevenly. In other words, the points where boundary layer separates from the ball's surface keep shifting. This fluctuating separation points create uneven pressure around the ball, producing unstable forces that can push it in different and altogether random directions. As a result, the ball may dip, swerve, wobble, or veer off wildly, following no clear path, as though it has a mind of its own. This crazy spectacle is known as the *knuckleball effect*.

And Ronaldo had mastered this craziness. By striking the ball with immense power but absolutely no spin, he triggered the knuckleball effect that made the path nearly impossible to predict for the keeper. The ball could have deviated just wide or dipped too soon, yet it tore right into the net that day.

Two years later, in the 2010 FIFA World Cup held in South Africa, Adidas introduced its boldest creation: *the Jabulani*. Hailed as the "roundest ball ever" with only eight seamless panels and aerodynamic grooves, jabulani took the knuckle ball madness to a whole new level. It soon became keeper's nightmare due to its unstable path. On top of that, many World Cup matches took place in high-altitude cities like Johannesburg, nearly 5,500 feet above sea level. Because of thinner air, drag itself gets reduced which means the ball traveled faster and swerved even more crazily. Its unpredictable movement drew so much attention that NASA scientists at the Ames Research Center got curious and decided to investigate on their own. They tested it in wind tunnels, just like they do with aircraft. They found that at speeds around 45 to 55 miles per hour, typical for a free kick, the Jabulani's ultra-smooth design exacerbated the knuckleball effect. The ball moved in a way that suggested it was dodging unseen forces mid-air.

So not exactly rocket science but it did require NASA and their wind tunnels to figure out what the Jabulani was up to. FIFA, quite sensibly, vowed not to use it again, and Adidas ended its production in 2012. Yet legend has it that the soul of the last Jabulani still roams out there—drifting, wobbling, eternally unsure of where it is going. By the way,

Adidas wasn't lying though; it really was the roundest ball ever. And if I were a goalkeeper, I'd have declared it the *most spherical 'individual'* I've ever encountered. I'll say no more. If you know, you know.

Anyway, the effects of spin or the lack of it aren't soccer-specific. Other sports feel them just as strongly. From the courts of tennis to the diamonds of baseball, the trajectory of a ball turns into a tool of mystery.

The Art of Dipping

In tennis, players use topspin to bend the ball's path. When a player applies topspin, the ball rotates along the horizontal (x-axis) as it travels. The spin causes the airflow on top of the ball to slow as the surface moves against the air while speeding it up underneath where the surface moves with the air. This creates higher pressure on top and lower pressure underneath, forcing the ball to dip sharply, clearing the net with ease and bouncing high off the court.

If there's one player who has turned this into a form of art, it is Rafael Nadal. Known as the King of Clay, Nadal's forehand generates a staggering spin of 3,200 revolutions per minute (RPM) or about 335 radians per second. Oh yes... No typo, you read that right. This exceptional spin rate is significantly higher than that of many of his peers. Roger Federer's forehand topspin, for instance, averages around 2,500 RPM, while players like Pete Sampras and Andre Agassi averaged approximately 1,800 RPM.

When Nadal delivers his topspin, the ball arcs over the net before dipping instantly, and after hitting the court bounces high to torment opponents. Facing these shots forces players into awkward positions, often striking at shoulder height or higher, which is an unenviable task to say the least. Nadal's topspin finds its ultimate expression on clay courts, where the slower surface exaggerates the ball's bounce, earning him a record-breaking number of French Open titles and cementing his dominance into the red dirt of Roland Garros. And on May 25, 2025, his legacy was cemented literally—his footprint, along with his name and the iconic 14, was permanently carved into Court Philippe-Chatrier.

The Smooth Slice

While topspin forces the ball to dip and bounce high, backspin or *the slice* does the opposite. By rotating the ball backward, backspin slows its descent, making it linger in the air and bounce lower upon landing. Few have wielded backspin as adeptly as Roger Federer. His backhand slice breaks the tempo of aggressive baseliners, neutralizing opponents on fast courts like grass and hard surfaces.

I don't know about other fans, but my love for tennis was born in the crucible of Nadal and Federer's rivalry. It was the rivalry of the purest kind where each pushed the other to unearth new depths of greatness. And yet off the court, there was respect, admiration, a form of unspoken friendship. I mean, how rare is it to witness a player weep at his rival's farewell? In 2022, as Roger Federer played his final match, a doubles game alongside Nadal at the Laver Cup, the world saw something extraordinary. Nadal laid bare his heart and broke down in tears.

Together with Djokovic and Murray, they formed a golden quartet that turned an entire era of tennis into the stuff of legends—each point a story, each match a drama. They showed us that rivalry doesn't have to mean enmity; it can also mean mutual elevation. When I think of those years, I don't just see forehands and backhands. I see heart. I see camaraderie. I see proof that sports is above petty fights and small egos.

The Tricks of Pitchers

In baseball, pitching is as much about physics as it is about finesse, and sometimes the line between them gets blurry. Let's begin with the *curveball* whose name says it all. It's a pitch that bends during its flight curving away from the batter's expectations. When a pitcher applies topspin, air pressure builds on the top of the ball while dropping underneath, forcing the ball to dip sharply as it nears the plate.

Then we have the slider, which combines elements of a fastball and curveball, featuring both lateral and downward movement. Its spin axis is tilted, resulting in a pitch that breaks diagonally with a subtle but late-breaking movement that often leaves batters swinging at thin air, fooled until it's too late.

When a ball is thrown without spin, or the knuckleball, small shifts in airflow around its seams create turbulence. As a result of this, ball wobbles and dips unpredictably, making it nearly impossible for batters or even for catchers to cleanly receive it. To the untrained eye, its erratic flight might look like a mistake. In reality, throwing a good knuckleball is a skill that demands precision.

A longtime standout for the Boston Red Sox, Tim Wakefield confused batters for decades. Sometimes his floating pitch would flutter several inches in random directions before crossing the plate. R.A. Dickey was another famous pitcher who made it his primary weapon. Unlike traditional knuckleballers, Dickey's version was faster, combining the chaotic motion of a classic knuckleball with the speed of a conventional pitch, leaving batters baffled and spectators stunned.

Some of the greatest breakthroughs in human history didn't come from rigid calculations and structured experiments, but simply by paying attention. There was no manual explaining that a scratched golf ball would travel farther than a pristine one, neither did any scientist proclaim it from a podium. It was the players themselves who took note of it, those who played the game with open eyes. That observation led to one of the most important aerodynamic revelations. Decades later, similar story happened on cricket grounds. Reverse swing, the supposedly supernatural ability to make an old ball swing in the opposite direction of what was the norm, wasn't born in a physics lab. It was a quirk noticed by keen-eyed bowlers. They observed how an aging ball, under specific conditions, behaved differently. What followed was a better understanding of turbulent airflow and a groundbreaking strategy in the game of cricket.

And this does not stop on the sports arena. Alexander Fleming didn't set out to cure bacterial infections. He saw something peculiar on a forgotten petri dish that a mold had killed the surrounding bacteria. What may have been a contamination to others was a *Eureka moment* to him. That single flash of awareness changed the course of medicine and eased human suffering by leading to the discovery of penicillin, the world's first antibiotic.

Could the lesson be any simpler? Open your eyes wide, your minds even wider. Don't stop at seeing what it is, see what it could be. Isn't it surprising how easily we overlook the power of observation? An ability that demands no advanced degrees or years of training, only the willingness to notice what others ignore. However, caught in the perpetual race of life—competing, hasting, striving—we all too often miss the signs the world constantly offers. We run after complexity, believing that the profound must be buried beneath layers of difficulty. Yet in reality, the answers lie in plain sight if only we see. Though the caveat is that eyes should be untangled from expectations and mind free from preconceived notions.

It also brings to light the truth that, in objective terms, nothing is by its very nature a flaw. Sometimes, a mere shift in perspective is enough to change everything. For it is precisely these so-called defects in a sports ball that transform sport into art and players into artists. A flawless golf ball wouldn't reach as far. A smooth cricket ball wouldn't curve through the air in defiance of expectation. All of it comes not from perfection, but from what appears imperfect.

So, what about our personal imperfections? Couldn't it be that our so-called flaws are something we have yet to recognize? We spend so much time hunting exactitude, sanding down our rough edges, trying to fit some imagined ideal, when perhaps it is the dents we bear, the scars we hold, and the unevenness we carry that tell the most honest story of who we are. The paintings of our lives are never set up with neat brushes on a flawless canvas, instead they are filled with unexpected strokes.

We admire smoothness but doesn't a smooth boundary layer detach too soon, leaving behind a large chaotic wake that slows progress? In contrast, a rough surface disrupts the flow just enough to create turbulence within the boundary layer which then propels the object farther. Have calm waters ever made skilled sailors, or smooth highways, seasoned drivers? If ease breeds nothing, so why assume a struggle-free life will take us somewhere worth going? Our rough patches are our moments of turbulence that push us forward. The smoothest path may look inviting, but it is the rough one that takes us the farthest. Again, does a dented golf ball fall short or does it fly twice as far?

Finally, if life ever feels uninspired, confined to a rigid path, why not add your own twist? Why can't there be a Magnus effect in the linearity of life? Why keep on marching the straight road that numbs the soul? In the end, it is the bold and breathtaking trajectory that stays in memory—whether in sports or in life.

9

The Poetry of Fluids

It's a private art collection in New York. The room is filled with paintings, each of them vying for the spotlight and asserting its presence individually. From vivid landscapes to striking portraits and from abstract patterns to lifelike figures. The warmth of light glowing over the canvases is accentuating the textures of thick brush-strokes. The faint scent of varnish, mingled with the aroma of wood, lingers in the air. Everyone is immersed, some in their thoughts, others in the exquisiteness of color and form.

Amidst the gallery, one canvas is standing apart, as if impossible to ignore. From across the hall, it is pulling you, drawing you in before you even understand why. It isn't politely inviting your gaze, nor is it flirting for attention or asking for admiration. It is, in fact, demanding it... All of it. There is reverence, and there is awe. And then there is something more—something harder to name. Something that stirs in your chest when you find yourself too close to beauty, and your heart begins to throb like a bird fluttering helplessly against the bars of its cage. That something...

Up close, the utter scale of it swallows one whole, or so it seems. There is no focal point, no clear beginning, no definite end. The colors—yellows, browns, grays, and whites—are colliding, flowing, jostling, and jolting. If you look, it looks noise. But if you feel it, only then do you see. And the longer you feel, the more the patterns emerge. Fluid curves guide the eyes in spirals and arcs as the streaks aren't random at all.

Surprisingly, no brush was used to create it. The painter didn't sit at an easel to paint it with careful strokes. He laid the canvas flat on the floor, and circled it like a dancer. Using only cans of paint, he let the paint splatter and drip, cherishing the unpredictability of flow, and speak its own language. He captured the poetry of fluids in its purest form and let the paint compose its own verses as gravity pulled it, viscosity dictated its thickness, surface tension shaped its edges, and momentum carried it further to complete the poem.

This ode to fluid mechanics is "No. 5, 1948" by Jackson Pollock that shattered records when it was sold for an astounding $140 million in 2006, the highest price ever paid for a painting at the time.

Some see it as a reflection of Pollock's inner unrest. Others find it an expression of freedom, rebellion, creation, and destruction, or perhaps all at once. Whatever one may see, one thing is certain that Pollock let his emotions flow through the motion of paint, and from the forces that stir storms, he conjured one of his own on canvas.

We don't know if it was Pollock guiding the motion of the paint, or the motion guiding him? It also leaves us wondering between the boundaries of art and science. Do we even know where does one end, and where does the other begin?

Although Pollock's *drip paintings* exemplifies how fluid mechanics connects with art, it isn't confined to a single technique or a particular medium. The behavior of paint, governed by properties like viscosity and surface tension, runs into every stroke and all artistic styles. Let's explore how these fluid properties bring paintings to life...

Viscosity and its Role in Art

When paint spreads across a canvas, what really governs its movement is viscosity. Remember we discussed it back in Chapter 3? It is simply a measure of how thick or thin a fluid is, or how resistant it is to flow. One can consider it like a fluid's personality—thick and slow like honey, or thin and free-flowing like water. The bold strokes and swirling colors may steal the spotlight, but it is viscosity that dictates the flow in nuanced ways. In the hands of an artist, viscosity finds a voice capable of stirring the entire range of human emotion.

A great example of this is *fluid acrylic pouring*. In this exciting style, artists work with thin, low-viscosity acrylic paints that flow freely across the canvas. By scrupulously thinning the paint with water or a specialized pouring medium (a liquid to enhance flow without diluting color) painters achieve a consistency that allows the paint to spread easily. Colors mingle together, forming stunning cells that resemble intricate microscopic patterns. What takes shape is a vibrant display of colors, entirely carried out by the fluid's ability to flow. Paintings like Holton Rower's *Pour* series, with their cascading waves of color, or Emma Lindström's fluid works capture the magic of this technique. If you have never seen fluid acrylic pouring in action, it is definitely worth a look. It gets far too easy to lose oneself in and one video turns into five.

In contrast to this technique is *traditional oil painting* where the paint's high viscosity allows for the creation of texture and structure. The term oil painting comes from the medium itself where pigments are suspended in oil, typically linseed oil, which gives the paint its thick consistency. This oil base gives the paint its liveliness and also extends its drying time considerably. This allows artists the freedom to rework and refine their compositions over days or even weeks. Smooth, buttery strokes stick to the canvas, holding their shape with a tactile quality that gives a sculptural vibe to it.

With this exclusive property, artists can build layers and add dimension or physical depth to their work. The raised ridges of paint catch and reflect light, highlighting the contrast between shadow and color. These textures seem to beg for touch, as if asking for a closer connection. Artists like Vincent van Gogh famously used the qualities of oil paint, layering it with bold strokes. He exhibited it to perfection in *The Starry Night*, where the texture of the paint conveys an emotional force that draws the onlooker into the scene.

Likewise, in *resin art*, where liquid resin is poured onto a surface, the thickness shapes the final piece. Thicker resin holds its form, preserving crisp lines and structured details, while thinner resin allows colors to flow candidly, blending into fluid patterns that match the natural movement of water or air.

Capillary Action and Surface Tension

You have probably never given too much of a thought to why water climbs up a paper towel or why raindrops like to keep themselves so neatly round in shape. We are so used to seeing them that they feel too normal to need any explanation, right? Both of these, along with a host of many more in nature, are driven by *capillary action* and *surface tension*. First let's get familiar with them and then explore how they influence art.

These ideas rest upon two love stories: cohesion and adhesion. Cohesion is the attraction between molecules of the same kind, while adhesion is the attraction between different kinds of molecules. For example, when water comes into contact with a solid surface, it may stick to it like glue if the adhesive force between the water and that solid is stronger than the cohesive force holding the water molecules together. In other words, cohesion is water's loyalty to itself and adhesion is its fascination toward other things. If the surface is more inviting than its own kind, water leaves the group and sticks to the surface. Which of them wins, whether cohesion or adhesion, depends on what the liquid comes into contact with. Let's take a look at how they show up around us.

Dip the corner of a paper towel into a spilled drink, and you'll see the liquid climb up the fibers on its own. Since the water molecules are strongly attracted to the solid fibers of the paper towel, this adhesive force is strong enough to overcome the cohesive pull between the water molecules themselves, drawing the liquid upward through the tiny gaps. This process is called *capillary action*. It happens when adhesion (the attraction between a liquid and a solid surface) takes the lead, and the liquid can move up through tiny spaces without any external push. The same process takes place in nature when plant roots pull water from the soil, and lift it against gravity.

Next, we have *surface tension*. Picture a pool of water. Deep inside, every water molecule is surrounded by others just like it. They sit comfortably in balance as they are pulled in equally from all directions. Now think about the layer of molecules at the surface of the water. This layer of water molecules is in a tricky situation because it has two very different neighbors. Beneath it are fellow water molecules which is lovely company, one of the same kind. However, above it is bland air, for which it doesn't find any attraction. There is no spark, no connection. So, what should this top layer of water do? Since water molecules are attracted more to each other than to air, the surface molecules are pulled inward (and sideways) by their watery neighbors. This inward and side pull creates a kind of tension at the surface like a stretched skin, which holds the water together, and so we have the name surface tension.

If you want to see it in real time then try filling a glass of water right to the top. Now, very slowly add a few more drops. Instead of spilling, the water rises above the rim and forms a rounded bump. The thing that holds the water together from spilling is surface tension. Obviously if you add too much, gravity takes over and the water spills.

Alright then, why are raindrops shaped like spheres? A raindrop is surrounded by air from outside, but inside the water molecules are pulling on each other. Surface tension pulls everything inward, trying to keep the drop as tight and small as possible. The natural shape that comes out of it is neither a cube, nor a cone... Yes, you guessed it right. It is a round shape. A *round* of applause for you.

You don't seem too convinced, are you? No problem, let me give you another classic example. And unlike those WWE warnings, this one you *can* and absolutely *should* try at home. Take a needle and gently place it on the surface of water. Even though the needle is much heavier and denser than water, it won't sink. Because surface tension creates a kind of thin, elastic skin on the water's surface. This invisible skin or membrane is strong enough to hold the needle in place, preventing the surface from breaking. Yes, there's some adhesion between the needle and the water, but it's not enough to drag it under. The water molecules attach more tightly to each other than to the needle, preserving the surface tension and keeping it afloat. We say the needle is *hydrophobic*, which literally means water-fearing, the opposite of which is *hydrophilic* or water-loving. Water, please don't take the needle's rejection personally. You'll find love elsewhere. Hmm... like in paper towels.

In nature, the same phenomenon allows insects like water striders to walk on water without sinking, since surface tension is strong enough to support their weight, and their legs don't disrupt the water's cohesive forces.

Okay, one last example. Yes, yes, sure. I promise. If you have ever tried blowing bubbles with plain water, you know it's a lost cause, don't you? The bubble pops almost immediately, if it even forms at all. As water's surface tension is strong, this strong cohesion pulls the surface molecules inward and makes it hard for the water to stretch and form a bubble. The surface just snaps back, popping the bubble before it can hold its shape. Now if we add a little soap, we can watch how water changes its allegiance. Soap weakens the cohesive forces between water molecules. With less cohesion, the surface tension drops, and the water becomes more flexible and easier to stretch. For an analogy, you can picture it like a crowded gym class. Without soap, the water molecules are like people standing shoulder to shoulder. With arms crossed, nobody is moving, let alone stretching. But when soap walks in like a strict yoga instructor asking everyone to give each other a little space and telling them to loosen up. Instantly, the whole group relaxes and there's room to stretch.

To sum it up, capillary action and surface tension are both governed by forces of cohesion and adhesion. In capillary action, adhesion takes the charge and can pull liquid upward through tiny spaces. In surface tension, it is cohesion that dominates, forming a tight surface that holds the liquid together.

Now, let's return to the realm of art. In *watercolor painting*, for example, as water seeps into the tiny fibers of the paper, it pulls the pigment along with it. Capillary action spreads colors to create soft and organic gradients that give watercolor painting its natural charm. This movement happens because the paper's fibers attract water molecules with enough force to overcome their natural pull toward each other. The pigment-laden water travels through the narrow channels within the paper and capillary action itself becomes an active participant in the art.

Meanwhile, surface tension operates on a different level depending on the canvas. On dry paper, surface tension holds the liquid together, encouraging droplets to bead up and resist spreading. This creates precise outlines and sharp details, especially on less absorbent or heavily sized paper. At the same time, capillary action keeps pulling inward, gradually breaking the droplet apart as it's absorbed. On a wet paper however, surface tension weakens at the droplet's edge because now water merges with the existing moisture. The resistance fades, allowing pigment to bloom outward in seamless transitions, making spreading easier and more fluid. Hence many artists intentionally pre-wet their paper to lower surface tension for better fluid blending. This preparation softens harsh edges, evens the distribution of pigment, and enhances the natural flow of color. By adjusting the amount of water in their brush or on the paper, artists control how colors merge or stay defined.

The beauty of these two forces isn't limited to water-color painting alone and their influence extends to many other forms. In ink art, for instance, capillary action pulls ink into branching patterns reminiscent of tree roots or veins. In sumi-e ink art, the movement of black ink across rice paper creates almost meditative lines. Similarly, in alcohol ink art, where pigments are mixed with isopropyl alcohol, surface tension creates striking shapes and as the liquid evaporates, it leaves behind flowing trails of color.

It was 12th-century Japan. A Buddhist monk was sitting at the edge of a pond. He was holding a brush in hand and a thought in mind. For the monks of this age, calligraphy was meditation, and painting, a path to spirituality. These were not isolated acts of artistic expression, but sacred rituals—mirroring the outside world and the one inside.

As he was preparing to transfer his thoughts onto the canvas, the bottle of ink slipped into the water. Much to his surprise, the ink didn't dissolve. Instead, it spread across the surface of the pond and unfurled into exquisite patterns. Intrigued, the monk began to experiment. He dripped ink more deliberately and watched it float with a mystical grace. With the gentle tip of a stick, he guided the ink into mesmerizing designs. The patterns shaped by fluid motion offered a sense of inner calm, as though suggesting that science and spirituality had always been one. This monk may not have realized it, but he was practicing the earliest form of an art what we now call *marbling*, or *suminagashi*, which translates to floating ink.

What is thought to have begun by mere chance soon evolved into intentional artistry. Centuries later, the art journeyed westward to Turkey where it progressed into ebru or cloud art. Turkish artists added their own innovations by using combs, needles, and even horsehair brushes to create designs. The term marbling originates from its resemblance to the vein-like patterns found in natural marble. It is an art form that feels almost ethereal. Unlike traditional paintings, where the artist's intent and precise brushstrokes define the image, marbling allows the liquid to paint itself in whatever way it wants.

The process begins by gently floating the pigments on water or a viscous liquid like carrageenan (a thickening agent made from seaweed). The pigments don't mix with the liquid below and spread across its surface in thin layers. Surface tension has the starring role. Differences in surface tension cause pigments to expand into fluid and organic shapes. Artists then manipulate these patterns, gently blowing on the surface, drawing combs through the colors, or using their fingertips to guide the motion with each movement reshaping the patterns. When the design is ready, a sheet of paper or fabric is gently laid onto the surface of the water. The paper's fibers or the fabric's texture absorb the floating pigments via capillary action, capturing the patterns exactly as they appear on the liquid. The process demands a steady hand and an immaculate timing, as even the slightest ripple can shift the entire composition. Once the transfer is complete, the paper or fabric is carefully lifted and a stunning, one-of-a-kind print is ready.

When I first saw marbling—just recently, to be honest— I wasn't prepared for how hypnotic it would be. It felt as if capillarity and surface tension had picked up a brush to paint the ephemeral beauty of fluid motion, and then froze it forever in a lasting image. In this age of digital abundance, you have really no excuse not to see it for yourself. You will find it far more enjoyable than most of what we scroll past in the name of "content."

Isn't it fascinating to realize that art and nature are painted with the same brush of principles? Doesn't it stir something deep within that the ascent of water through a plant's roots shares the same phenomena that give a Van Gogh painting its soul? As Pollock said,

> *When I am in my painting, I'm not aware of what I'm doing... I have no fears about making changes, destroying the image, etc., because the painting has a life of its own.*

Indeed, his paintings did have a life of their own, vividly alive yet so inanimate, and born from the properties of fluid itself. And while breathing life into a canvas, fluids leave behind a lasting lesson for our lives as well.

If fluids rely on cohesion to hold themselves together, shouldn't we, too, possess an inner cohesive force to bind ourselves? Without our values and our principles, what holds us together? If there is no sense of right and wrong left within us, isn't our soul left scattered then?

Although bound by its own cohesion, a fluid still needs something beyond itself. It must find another surface to hold onto if it is to ascend higher, high enough to make even gravity bow before it. Isn't it only natural, then, that we, shaped by the same nature, must have adhesive qualities too?

Nevertheless just as a fluid doesn't stick to any random surface, neither should we. What we choose to follow should arise from conscious awareness, and not from the blind pull of our DNA. Ask yourself before getting attached to an ideology, cause, or even a pair of charming eyes: *Will this attachment elevate me in any way, shape, or form?*

If you find the answer in affirmative, go ahead. If not, turn a *hydrophobic* eye and walk away, because this is what nature reveals and this is what art reflects.

10

The Architect of Modern Living

It's one of your routine mornings. The alarm buzzes and pulls you from a comfortable night's sleep. The room is nicely cool, owing to air conditioning. You shuffle to the bathroom and turn on the switch that instantly illuminates the space. The electricity, behind the switch, is generated by turbines spinning miles away. You tap the shower, and a jet of warm water rushes to greet you, delivered via pumps.

Fully awake now, you make your way to the kitchen. The coffee machine hisses, forcing hot water through finely ground beans at just the right pressure to brew a jolting cup. Mug in hand, you step into your car. A turn of the key brings the engine to life by mixing air with fuel in a controlled explosion. As you drive, traffic lights flicker ahead, powered by the same turbines that lit up your bathroom moments ago. The car's air conditioner hums, pushing air through vents.

This routine feels so ordinary that one never thinks about the extraordinary science that makes it all possible. It gets overlooked rather easily. So, let's pull back the curtain and see how the applications of fluid mechanics drive our daily lives.

Picture yourself waking up each morning and trekking miles upon miles just to fetch enough water for the rest of the day. For much of human history, people weren't merely picturing it; it was their grueling reality.

Now, imagine telling someone from 500 years ago that in the future, no one would need to haul buckets from a well or river. Instead, water would flow directly into the homes, ready for drinking, cooking, washing, bathing, gardening, and whatnot, all without lifting a finger... okay, maybe just a little nudge of it! Or tell them that no more chopping wood or burning oil to light up the night or warm the room. Instead, entire homes would glow by an invisible force... again, with a gentle tap of the finger. How do you think they'd react?

There would be some who might burst out laughing. Others would call it sorcery and accuse you of spinning the wildest fairy tale they have ever heard. In my case, people would've just stared me in utter silence, unsure whether to throw me out or fetch me some medicine. And yet, here we are, living in a world what would have been a fantasy to many.

Let's start with the water flowing into our houses. Water and the next section is brought to you by .... pumps. *Tragic oversight, I must say. Dear pump manufacturers, it's never too late.*

Pumps: Liquid Movers

A pump is like a transporter, moving liquids from one place to another. In homes, it provides water from reservoirs to taps. In cars, it pushes fuel to the engine. In industries, it moves everything from oil in pipelines to chocolate in food factories.

I trust you haven't forgotten that fluids flow from regions of high pressure to low. Pumps take full advantage of this principle. They create regions of high pressure, pushing liquids through pipes toward low pressure areas. If you want, you can visualize them as giving the fluid a small push or, depending on the application, a mighty shove. There are two major types of pumps, each with its own unique way of moving fluids.

Centrifugal pumps

Picture yourself on a merry-go-round. As it spins faster, you feel yourself being pulled outward, right? Imagine if there's no seatbelt, no handles, no safety rail. Perhaps best not to imagine it, really. Centrifugal pumps are, quite simply, a merry-go-round for fluids. Inside, they have an impeller—a component similar to a fan blade—that rotates rapidly and transfers energy to the fluid. And since the fluid has no seatbelt or even a seat, it's flung outward due to centrifugal force, gaining speed as it spirals away.

But wait a second... aren't we trying to increase the *pressure* of the fluid to move it, so what is with all this flinging and throwing and speeding up? "Long time no see," says our old friend Bernoulli.

As the high-velocity fluid exits the impeller, it flows into a special chamber, either a diffuser or a volute. Some pumps employ diffuser, others use volute. The difference between them is only of design, otherwise their purpose is same, which is to convert the fluid's velocity into pressure. A diffuser guides the fluid through cleverly designed pathways, called *vanes*, that allows it to spread out gradually or diffuse it. As the fluid spreads out, its velocity reduces and pressure increases. A volute, on the other hand, is a spiral-shaped chamber. As the fluid travels through the spiral of increasing area, its speed reduces naturally, and pressure builds up.

To understand, let's revisit the analogy from Chapter 6, where air flow was compared to traffic. Imagine cars cruising along a two-lane road until, without warning, the road narrows to a single lane. To avoid a traffic jam, every car *must* speed up to keep traffic flowing smoothly. Fluids are our ideal drivers—when they pass through a narrower section, their speed increases and their pressure drops. And the reverse is also true: when the road widens to four lanes, or a fluid moves into a broader channel of increasing area, its speed drops and its pressure increases.

Centrifugal pumps can move enormous volumes of liquid, which is why they are a backbone of modern infrastructure. They keep water moving through city pipelines, feed irrigation systems, and drive industrial operations.

Positive displacement pumps

As kids, most of us were terrified of injections, and by the way, I'm still not a huge fan. Anyway, whenever you received a shot, while you sat there valiantly pretending the sting was nothing more than a polite pinch, you were unknowingly witnessing the workings of a positive displacement pump.

I know, it is not the friendliest way to meet such a delightful concept of physics. That too demonstrated by a person with a needle in hand the size of a harpoon. When the doctor or nurse—both equally enthusiastic, mind you—press the plunger with the gleeful look of a seasoned torturer, the space inside the syringe shrinks and forces the liquid to move out in a stream straight into your poor, unsuspecting arm. Well, that is how a *piston pump*, which is a type of positive displacement pump, works. A piston moves inside a cylinder while squeezing the fluid forward with each stroke.

Positive displacement pumps operate by trapping a fixed volume of fluid and then pushing it forward, creating a steady flow. Unlike centrifugal pumps, which rely on increasing velocity and then converting it into pressure, positive displacement pumps work by directly increasing the pressure. The key principle here is to reduce the space available for the fluid, so that it finds no choice but to move. Inside such pumps, a mechanical component like a piston or gear creates a sealed chamber where fluid is trapped temporarily. As this component moves, the volume of the chamber shrinks, forcing the fluid out.

We have already seen one type i.e. the piston pump. Another common type is the gear pump. Consider two interlocking gears spinning inside a tight casing. As the gears rotate, fluid gets trapped in the spaces between the gear teeth and the casing, and then carried around to the discharge side.

Positive displacement pumps are used when a precise and consistent flow is required. For instance, medical devices like infusion pumps deliver exact amounts of medication at controlled rates. Many chemical industries use positive displacement pumps to handle viscous fluids. Likewise, in food and beverages lines where products like syrup, yogurt, or chocolate require gentle but consistent handling, these pumps are employed.

And while we are at it, guess what else qualifies for a positive displacement pump? Yes. The one inside our body. The heart might not have pistons or gears, but with every beat, it traps a fixed amount of blood and pushes it forward. Pumps don't look as glamorous as other tech devices—no AI involved, sorry—but they keep our flow just right. From the steamy shower that washes away the day's poor decisions to tiny jab delivering medicine to the blood coursing through our body to the chocolate we all love.

Compressors: Compressing the Gases

For gases, we have "special pumps" called compressors. Unlike liquids, gases can be compressed significantly because they have a lot of empty space between their molecules. As the name suggests, these devices work by taking in a gas and compress it into a smaller space to reduce its volume. The reduction of volume increases the pressure and eventually the gas is sent out. This nifty process makes them essential in all applications where high-pressure gases are needed, such as refrigerators, air conditioners, jet engines, industrial tools, medical oxygen tanks, and the list goes on. Like pumps, compressors come in two similar types as well, based on how they compress gases.

Positive displacement compressors work by trapping a fixed amount of gas in a chamber, squeezing it to reduce its volume, and meanwhile increase its pressure. Believe it or not, but chances are you have already used this type of compressor a gazillion times—a bicycle pump! Although we insist on calling it a pump, if we are being technically accurate, it's a compressor actually. Yes, calling it by its proper name "bicycle positive displacement compressor" might delight the engineers but confuse everyone else. It sounds more like a piece of equipment from a high-tech factory, and not something gathering dust in your garage next to an old toolbox. So let's be sensible and stick with the name pump as it is shorter and simpler, and far less likely to frighten children or scare a cat. As a matter of fact, no one will ever take you seriously again if you say, *Can you please pass me the positive displacement compressor, I've got a flat tire!*

Coming back to the point, every time you push down on the handle or plunger, you are moving a piston inside the long, circular body called the cylinder. This forces the air to squeeze into a smaller space, building pressure until it's high enough to inflate the tire. On a larger scale, a powered version of the same idea is behind refrigeration systems and many such high-pressure gas applications.

On the other hand, *dynamic compressors* work much like centrifugal pumps. They use rotating impellers to increase the velocity of the gas, then slow it down in a diffuser where that velocity is converted into pressure. As they can handle large volumes so it makes them ideal for large-scale applications like jet engines, power plants, and gas pipelines.

Now that we have seen how pumps and compressors work, it's about time we appreciate some applications that rely on them. Let's open the act with a scene that might just send a shiver down your spine...

A Cool Performance

Refrigerators in our kitchens are the ultimate food savers, or shall I say the mood savers too. To keep food fresh, they need to maintain low temperatures in order to slow down the reproduction rate of harmful bacteria.

Refrigerators work by circulating a *refrigerant* through coils (it's a fancy name for tubes) embedded in the inner walls of the fridge. A refrigerant is a special fluid that evaporates i.e. changes from a liquid state into a gas state at extremely low temperatures. For example, common refrigerants like R-134a boil at around -26°C (-15°F). For comparison, water boils at 100°C (212°F). This low boiling point allows refrigerants to absorb heat and evaporate even in cold environments, such as the inside of refrigerators, which is typically maintained around 2° to 4°C. The process of evaporation cools the surrounding area and creates the chilly conditions.

You can see it for yourself by placing a few drops of alcohol on your skin. As the alcohol evaporates, a cooling sensation will be felt. We can experience this with water too, but since alcohol has a lower boiling point, it works much faster. With water, the effect is more noticeable when we step out of a shower and sit under a fan. The water on our skin quickly evaporates, leaving behind a refreshing coolness as it pulls heat away from the skin.

To start the evaporation process in the refrigerator, the pressure of the liquid refrigerant needs to be reduced first. To achieve this, the high-pressure liquid refrigerant is released into the coils via a tiny outlet called *capillary tube*. As the high-pressure liquid refrigerant passes through this tiny outlet, its pressure drops to start absorbing heat.

It is similar when we use an air freshener or body spray. Inside the aerosol can, the liquid is stored under high pressure. The nozzle of the spray releases it into the open air. As the liquid exits into the air which is a lower-pressure environment as compared to the aerosol can, it quickly turns into a gas, spreading the fragrance.* The same happens inside the refrigerator in the coils where liquid refrigerant is released. These coils, aptly named *evaporator*, allow the refrigerant to absorb heat from the fridge's interior and gets evaporated while cooling the surrounding area. You must have noticed the slightly raised surface on the inside walls of the fridge, right? They are the evaporator coils.

---

*This is why aerosol cans carry warnings not to pierce or burn them. Even after the liquid seems used up, they often contain pressurized gas. Piercing it could cause a sudden release of gas, spraying leftover contents into your face. Not something you'd want, I tell you.

Okay, so the refrigerant has now cooled the fridge. Great! Thank you very much. See you later. Well... not quite. This isn't a one-time act where the refrigerant drops the mic and walks off stage. To keep the show running, the now-gaseous refrigerant needs to be turned back into liquid state. Here's where we need the *compressor* to compress the refrigerant gas, reduce its volume, and raise its pressure. As the gases get squeezed, their molecules are pushed closer together. Since liquids have much less space between their molecules, the gas usually condenses into a liquid when the pressure becomes high enough.

However, we have got a problem. The compressor has done its job, which is to raise the pressure high enough and yet, the refrigerant still hasn't condensed into liquid. The thing is as pressure goes up during compression, so does the temperature of gas. And high temperature makes it hard—nearly impossible—for any gas to condense. If we go back to a bicycle pump, you might've observed the cylinder and the rubber tube of the pump getting hotter as you inflate the tire, right? It is because the compression causes the temperature of air to increase as well along-side the pressure. Likewise, in the refrigerator, when the compressor finishes its job though the refrigerant becomes highly pressurized, it also gets really hot... Too hot to condense. A hot refrigerant isn't going to do us any good. What we need is a refrigerant that is cold, liquid, and under high pressure—all at once.

This brings the *condenser* to make its entry in this icy drama. The series of coils you see on the rear side of a fridge are condenser coils which act as a heat exchanger.

The job of a condenser is to calm down the hot, high-pressure refrigerant gas. As the hot refrigerant gas flows through these coils, it releases the heat to the surrounding air. This is the reason why the back of fridge feels warm. As the refrigerant releases the heat, it cools itself down, and as soon as it cools down, it condenses into a liquid. Since the compressor keeps pushing more refrigerant into the system, it doesn't allow the refrigerant pressure to drop down and it stays under high pressure.

Now the refrigerant is back in its ideal state: cold, pressurized, liquid, and perfectly ready to evaporate again. And so the cycle goes round and round, like the wheels on the bus. *Oh Lord... it loops in my head even as I try to sleep. Thank you, kids.*

Okay, here's a question. We have seen the actors of this performance—refrigerant, evaporator, compressor, condenser—but who is running the show? Who makes sure everything stays on cue, without hiccups? It's time to meet the director in charge: the *thermostat*. Thermostat keeps an eye on the fridge's temperature at all times. It takes action as soon as things start to heat up. The moment it senses a rise in temperature, it sends a signal to the compressor to spring into action. Once the thermostat is satisfied that the temperature inside the fridge is back on track, it gives the compressor the green light to relax.

And the cold play continues, keeping my ice cream frozen and preserving your five-day-old leftovers. No, I'm not judging you, since God bless my mom—she's fully convinced refrigerators can freeze time itself, let alone food. Hence a container of biryani from last winter is still in there. You know, *just in case.*

Heaven's Breeze

Now that we've cracked the cold case of refrigeration, let's tackle its bigger sibling—air conditioning. It takes the same science, nearly the same components, and scales it up. First things first, where does the cool air come from?

Again, it all begins at the evaporator coil, located inside the indoor unit of a split AC system. The first step begins with high-pressure liquid refrigerant flowing through the *expansion valve* (a component similar to capillary tube). This outlet releases the liquid into the evaporator. As the refrigerant flows through the expansion valve, its pressure drops sharply, causing it to evaporate and draw heat from the warm air around the coils.

At this point, a fan, also called a *blower*, takes center stage. It plays a double role: first, it sucks warm air from the room through *return vents*—the openings where air is drawn into the system—and blows it over the evaporator coils, which are filled with the cold refrigerant. The refrigerant absorbs the heat from the air, leaving it cooler. Next, the same fan pushes this now-chilled air back into the room through *supply vents*—the slats where the cool air flows out—providing the heavenly breeze on a hot day. In split AC systems, the supply vents are visible at the front of the indoor unit, while the return vents are hidden just behind or below them.

What happens next to the refrigerant that has absorbed all the heat? This now-warmed refrigerant, which has turned into a gas, flows to the compressor which squeezes the refrigerant to increase its pressure (and temperature as well), turning it into a hot, high-pressure gas. From there, it's off to the condenser, typically located outside the room or building, hence called as outdoor unit. Acting as a heat exchanger, the condenser releases the heat absorbed by the refrigerant into the surrounding air. As the refrigerant loses heat, it cools down and condenses back into a high-pressure liquid. And the cycle repeats.

Modern ACs or inverter models can *invert* the script and work as heaters too! They rely on a component called a reversing valve which reverses the cycle by changing the direction of the refrigerant flow. Now, the refrigerant absorbs heat from the outside air even when it's cold, in fact especially when it's cold. After that the refrigerant gets compressed, which raises its temperature further. This hot, high-pressure refrigerant then flows to the indoor unit, where it releases heat into the room. The indoor coils act as a condenser now. After releasing the heat, the refrigerant cools down. It is then passed through an expansion valve which further lowers its pressure and temperature, making it even colder to absorb heat more easily. The refrigerant then heads back to the outdoor unit to pick up the heat where the outdoor coil now becomes the evaporator.

Inverter ACs are a part of what we refer to as HVAC, an acronym for Heating, Ventilation, and Air Conditioning. In complete HVAC systems like those in large buildings, ventilation units are added as well for fresh air to flow in and stale air to flow out, completing the package.

The pumps, the compressors, the fridges, or the ACs, they all have one thing in common—their hunger for electricity.

Without it, they are nothing more than dead boxes of metal and wires. If we chase electricity backward i.e. from the outlet to the grid and from the grid to most power plants, we'll eventually find ourselves standing before something called *turbine*.

Turbines: The Engine of Electricity

A turbine is like the reverse of a centrifugal pump. A centrifugal pump spins its blades to fling water outward to increase water's velocity, right? Now, hit rewind. Picture a fast-moving stream of fluid (water, steam, or air) rushing in and striking stationary blades. As the fluid hits the blade, the blades start to spin or rotate. The blades are attached to a shaft (a fancy name for metal rod) and when the blades spin, the shaft spins too. Well, that is it. That is a turbine.

The kinetic energy (speed) of the fluid gets converted into mechanical energy (rotation) of the shaft. The shaft is then attached to a generator, which converts mechanical energy (rotation) into electrical energy (electricity). No matter the type of power plant, this final step—spinning shaft to electricity generation—stays the same. What makes one power plant different from another is the method they use to spin the shaft. That's how we end up with names like nuclear, hydro, thermal, or wind as they all just spin it differently.

Hydroelectric Turbines

In hydroelectric turbines, a stream of water rushes down from a height through a dam. Its potential energy (stored due to its elevation) is transformed into kinetic energy.

This powerful flow of water strikes the turbine blades, causing them to spin, and eventually rotates the shaft which then moves the generator.

I remember from my childhood, there was an elderly uncle in our acquaintances. He was a farmer and whenever his crops would fail, he used to shake his head and say, *The water wasn't of any worth this year again. How can crops grow if the government sucks out all the electricity from water through dams.* God bless his sweet soul. I wish he were still around so I could try explaining that while a dam does harness the energy of flowing water to generate electricity, it doesn't extract anything *out* of it. The water doesn't emerge on the other side nutritionally drained. Turbines are not some cunning thief that robs the water of its calcium, iron, or the will to grow crops.

Steam Turbines

In traditional power plants, fossil fuels like coal or oil are burnt to generate heat. The heat is used to boil water which produces high pressure steam. The fast-moving steam is then directed at turbine blades which spins them with immense force to generate rotational energy of the shaft.

But when it comes to nuclear power plants, many think of it like a science fiction with green glowing reactors, flashing red lights, and perhaps a shadowy alien-looking figure too. In reality, things are far less exotic. A nuclear power plant is just a glorified tea kettle. Yes, there is fission reaction of uranium atoms but it is only to generate heat—well, lots of it—that boils the water to produce steam. Once the steam is produced, it is business as usual. A steam turbine kicks into action spinning away to generate electricity.

I realize this may offend my dear nuclear scientists, many of whom would be thinking, "Splendid. Why don't you bring the tea leaves when you visit a nuclear power plant next time?" So, in anticipation of that, I have decided to become a milk lover from today. No uranium, only calcium. Less radioactivity, more bone strength.

Wind Turbines

Wind turbines operate on a similar principle of converting the kinetic energy of moving air into electrical energy. The massive rotor blades of wind turbines are shaped like elongated airfoils. Much like an airplane wing, each blade is designed to generate lift i.e. the curved upper surface and flatter lower surface create a pressure difference as air moves across them. However, unlike airplane wings, which are meant to generate upward lift, wind turbine blades are mounted at an angle so that the lift force causes rotation around the hub. This rotation turns a low-speed shaft, which is connected to a gearbox that increases the rotational speed. The high-speed shaft from the gearbox then drives a generator, where electromagnetic induction converts the mechanical energy into electricity.

Now that we know what powers the outside world, let's turn to something that powers the light inside our foggy brains each morning? Of course, it's coffee... What else were you expecting, a cucumber?

The Coffee Mechanics

Every great cup of coffee starts with beans and ends with the science of fluids that brings flavor to it. If coffee brewing were a race, *espresso machines* would be the Usain Bolt of it—fast, furious, forceful. They deliver bold flavors in record time. These machines heat water to a controlled temperature, typically between 90°C and 96°C (194°F to 205°F). This sweet spot for temperature is important because if the water is too hot—closer to boiling—it can scorch the coffee grounds, leading to a burnt taste. And if it is too cold, it under-extracts the coffee, resulting in a sour and weak flavor.

Once the water is at the right temperature, a pump is used to produce a high pressure of about 9 atm. The pump forces the heated water into a confined space where the coffee grounds are tightly packed in a portafilter (a metal basket with a handle). The pressurized water is forced through the tightly packed coffee grounds. For perspective, the high pressure of water is equivalent to a force of fully inflated car tire pressing down on the tiny espresso puck. This immense pressure, alongside fine grind and tight tamping, extracts complex flavors, aromatic oils, and the velvety crema in just 30 seconds. Without this exactness of temperature and pressure, the espresso would lose its iconic intensity, leaving with a lackluster shot.

Drip coffee makers, on the other hand, are the marathon runners of the coffee world—unhurried, steady, focused on the long game. Unlike the high-pressure sprint of an espresso machine, these brewers let gravity take charge, allowing hot water to flow gently over the coffee grounds. It soaks them evenly so that every bit of coffee gets its moment to shine. This helps avoid over-extraction or under-extraction, both of which can upset the balance of flavor. The aromatic compounds are gradually extracted, yielding a larger and milder cup which is ideal for those patient souls who relish taking their time with leisurely sipping.

It pairs beautifully with a good book, preferably one that talks about science... especially physics... specifically fluid mechanics. No, I'm not referring to *this* one. I am not that shameless. But if that's what you thought... well, I love you too.

Finally, we have the minimalist of coffee world—*the French press.* It begins with coarsely ground coffee and water having just the right temperature. Once the two meet, they are left to steep for several minutes in a slow infusion with no rush. When the steeping is done, the plunger, with its mesh filter, presses the grounds to the bottom, separating them from the liquid. Unlike espresso or drip coffee, the French press relies on diffusion, letting the water slowly pull out the coffee's rich essence.

For all its simplicity, there are certain things that must be just right. For example, grind the coffee too fine, and it will dissolve into water, leaving a muddy mess. Grind too coarse, and the water won't extract enough flavor, leaving a weak brew. Too hot a water will scorch the coffee, giving it a bitter taste. Too cold, and it will leave an uninviting cup. Steep for too long, and the flavors become harsh. Too short, and you'll get a brew that's underwhelming. *Did I mention minimalism?*

Yes. This is the bare minimum pain one must endure to earn a rightful cup of coffee. Lower the bar any further, and we'd be left with *instant* sadness that dissolves faster than our collective will for life on a Monday morning. That poor blend doesn't so much wake you up as reassure that staying in bed was the wiser choice.

11

A Tale of Betrayal

High above the earth, in the womb of a cloud, a droplet is waiting for its fall...

Its journey began when it was lifted from the ocean's surface by the warmth of the sun. For a while, the tiny vapor remained unknown and wandered alone, until it found others. Fragile on their own, yet they formed something greater than the sum of their parts. As the cloud wandered across the sky, it grew heavy with the weight of its gathered kin. And then, gravity called it down.

The real adventure began the moment it landed. It rolled over leaves, seeped into soil, and then emerged once more, slipping into a small trickle that wound its way downhill. That small trickle soon became a stream. It was no longer alone. It now belonged to something unstoppable.

Flow

The current carried it onward. It shaped the land, carved the valleys, and cut the mountains. It nourished life along its way in fertile fields where farmers relied on its presence. It plunged down waterfalls, meandered through bends, and then, after days, months, perhaps even years, it reached the ocean. It was a homecoming of sorts but only to be drawn by the sun again. Only to begin the cycle anew.

I find a haunting beauty, a strange and silent poetry, in this eternal loop that has been repeating itself for billions of years.

What appears to be the mere fall of rain is, in essence, the opening salvo of a grand process that molds the world. Each small droplet is an agent of change. Soft as they may seem, these tiny droplets have it in them what moves the mountains.

Where a raindrop lands is only the first decision in a long and unpredictable journey. Some are absorbed into the soil, vanishing beneath the surface to replenish underground reservoirs known as *aquifers*. Here, hidden from our sight, they nourish springs and feed wells that quench entire civilizations. Some drops hit bare soil like a small hammer, dislodging particles of ground and setting them in motion. Others strike impermeable surfaces like rock, asphalt, concrete where they have no choice but to keep moving. The energy of the water determines the rate of change. A heavy downpour displaces more soil, accelerating the process. They join forces to form rivulets that grow in strength as they race downhill. This surface flow is known as *runoff*. It creates streams that carry sediment and debris which cut small channels into the earth, insignificant at first but over time rills deepen into gullies and gullies into valleys. This is *erosion* that sculpts the landscapes.

The Mechanics of Erosion

Although erosion begins as soon as the water droplets from rain hit the surface, with rivers the phenomenon intensifies as it becomes a combination of several forces, all working together to break down and transport material.

Picture a river carrying pebbles and sand downstream. These particles act like tools, scraping and grinding against the riverbed and banks. With time, this *abrasive action* deepens the river's channel. Sometimes, water doesn't need these pebbles as it uses just its own force. A flowing river exerts pressure on cracks and crevices in rocks. The sheer force of the water pushes these cracks wider, causing pieces of rock to break off. This process, called *hydraulic action*, is especially powerful in fast-moving rivers. Finally, there's a third type of erosion. Rivers can dissolve minerals in the rocks they flow over, particularly in areas rich in limestone. As time goes by, this *chemical action* eats away at the rock, contributing to the river's ability to form landscapes.

One of the most vivid examples of river erosion is the birth of valleys and canyons. For example, the most awe-inspiring of them all is the Grand Canyon. Formed by the Colorado River, it stretches 446 kilometers (277 miles), reaching depths of more than a mile where each exposed layer of rock is a story in itself. Likewise, the River Thames in England may appear tame today but its ancient course formed the London Basin. Over millions of years, it carved through chalk and clay, leaving behind a broad floodplain and terraces that now support one of the world's great cities. The Nile River, flowing through northeastern Africa, is another example. Although celebrated for its role in enabling civilization, the Nile has long been defining the geography of Egypt and beyond through erosion.

The Mechanics of Deposition

Rivers don't just take away though, they also give back. As water flows, it picks up and transports sediment. The speed of river determines what types of sediment it can transport and how far it can carry them.

The fine particles such as sand, silt, and clay that the river can keep afloat in its current are usually referred to as *suspended load*. When the flow is strong enough, these fine sediments are suspended in the water giving rivers their characteristic muddy appearance. Larger particles such as pebbles, rocks, and even boulders move along the riverbed, and hence called the *bed load*. These materials don't float rather roll, slide, or bounce along the riverbed. This process is called saltation, and it requires a strong current to move heavier materials. In fast-moving rivers, bed load can be carried over long distances. Rivers also carry dissolved minerals, like calcium or sodium, that have been chemically eroded from the rocks the river flows over, called the *dissolved load*.

As a river slows down, its ability to carry the sediments diminishes. This is when *deposition*—or the dropping off of material—occurs. The largest particles are dropped first, followed by smaller ones as the river continues to slow. The deposition can create bars and sandbanks along its banks. When a river reaches its mouth and flows into a larger body of water, like the sea or an ocean, the current slows substantially, and the river deposits all the sediment it has carried. This creates a delta or a fan-shaped landform made of accumulated sediments. For example, the Nile River has been depositing sediment at its delta for thousands of years, creating one of the most fertile regions on Earth. Ancient Egyptian civilization thrived here, relying on the annual floods to replenish the soil with nutrient-rich silt. Similarly, the Ganges Delta in India and Bangladesh supports millions of people and numerous ecosystems.

However, not all rivers form deltas. For a delta to take shape, a river must slow down significantly and have space to deposit the sediment it carries. If the river's mouth is too inclined, the tides too strong, or the sea too deep, the sediments get swept away before they can settle and build up. The Amazon River, for instance, despite being one of the largest rivers in the world doesn't form a classic delta. Its enormous flow, combined with the Atlantic's strong tides and currents, prevents sediment from settling at its mouth. So, most of the sediment is carried far into the ocean, where it forms a massive underwater deposit known as a submarine fan.

In some other cases, rivers don't make it to the sea or ocean at all. Instead, they vanish into deserts or empty into inland lakes and basins. One such example is the Okavango River in southern Africa. Rather than reaching the sea, it fans out into the Okavango Delta, an inland oasis in the Kalahari Desert. The river's water simply spreads into floodplains and swamps to nourish one of the most biodiverse ecosystems on the continent.

Centuries before the Pharaohs of Egypt raised their pyramids toward the heavens, a civilization of equal grandeur was flourishing across South Asia. The *Indus Valley civilization*, one of the world's earliest urban societies, thrived from 3300 to 1300 BCE. It left behind traces of its brilliance in cities like Mohenjo-Daro and Harappa (modern-day Pakistan) that featured grid-like streets, advanced drainage systems, and purpose-built houses with massive granaries to store food for a growing population.

The Harappans built the infrastructure according to the flow of rivers—the Indus in the west and the Ghaggar-Hakra in the east. They harnessed the seasonal floods and sustained an agricultural system that fed their cities. For centuries, rivers and cities moved hand in hand, as lovers do. The water flowed, and the people flourished. But like all great love stories, sorrow found its way into this one as well. One of the rivers turned away, shattering the bond between land and water. Perhaps their love was too intimate not to draw the evil eye which spares nothing, least of all happiness.

The river, which had once held the land and its people close, eventually receded, abandoning the civilization it had so faithfully sustained. Its departure remains one of history's greatest mysteries. Was it the restless shifting of the Earth's crust? A creeping change in climate? Or something more inexplicable?

We don't know... Nobody knows...

What we do know, however, is that around 1900 BCE, the Ghaggar-Hakra, believed to be the fabled Sarasvati River, changed its course and began to dry up, shrinking into a seasonal trickle. And with it, fields turned barren, trade routes collapsed, and cities fell silent.

Some say the people migrated eastward, toward the plains of the Ganges River, where water still flowed. But again, no one knows for certain. Even if they did migrate, the Harappans, as they were, disappeared and their distinct urban culture vanished with the river. Streets that were once filled with trade are now buried in dust. Wells that once brimmed with water now stand forgotten.

The river changed its mind like a lover who suddenly grows distant and then walks away leaving behind no explanations—only questions, the scars of absence, and the lingering ache of abandonment. For the Harappans, those scars are still visible in earth's heart. All that remains are ruins and the enduring mystery of the betrayal.

Why Rivers Change their *Mind*?

The story of the Indus Valley Civilization underscores the fact that rivers are always in an endless pursuit of new paths. They are rarely content with moving in a straight line, and their shifting course creates the geography and sometimes decimates entire civilizations. On the outer edge of a bending river water flows the fastest and the river gains energy that scours the land, carving deep channels. On the inner edge the water slows down and here the river deposits sediment that builds up the land. Gradually, this push-and-pull causes the river to form arcs and curves, the familiar snake-like patterns known as meanders.

Meanders, though, are only temporary. As the bends grow more exaggerated, the river begins to coil back on itself. And when it happens, the river does what rivers do the best—it finds a quicker way forward by cutting a new shortcut and leaves behind its old path. What remains is an *oxbow lake*: a crescent-shaped shadow of the river's former self, now silently disconnected from the main flow.

Although erosion and deposition drive this process slowly, other forces can suddenly steer rivers off course. Tectonic movements like the jolt of an earthquake or the slow shifting of Earth's crust can tilt the landscape, forcing rivers to abandon their familiar routes and making entirely new ones. A change in climate, too, whether a prolonged drought or a season of relentless rain, can make rivers shrink back or surge outward. During a major flood, the water rises with such force that it may burst through levees and spill across plains. In the aftermath, the river may never return to its original course. Instead, it settles into the new path it formed in a moment of chaos. This sudden leap is called an *avulsion*, and it can redraw maps overnight, turning what was once dry land into a new riverbed.

In recent history, a startling example of avulsion took place on 18 August 2008 when the Kosi River—a tributary of the Ganges and known as the *Sorrow of Bihar*—broke free from its artificial confines. For decades, embankments near the Indo-Nepal border had held the river in place, but the eastern embankment at Kusaha in Nepal gave way on that fateful day. What followed was a nightmare.

The river changed its path and nearly 80–85% of the flow was diverted 120 kilometers eastward into an ancient channel where Kosi once used to flow over a century ago. Scientists suggest that this massive shift was triggered by years of sediment buildup within the embanked channel, which had raised the riverbed 4–5 meters above the surrounding floodplain. Villages that had existed far from the river's modern course, more than a hundred kilometers away, found themselves underwater in no time. These were places where no one had seen the Kosi in generations. Entire settlements disappeared as the river had quite literally moved in.

Unlike typical floods that drain into larger rivers, this new path didn't reconnect with the old Kosi channel, nor did it find proper through-drainage into the Ganges. More than 2.5 million people were affected, and the true toll on lives and livelihoods has never been fully measured. Something unspoken pulled the Kosi back that morning, stirred by a love now entombed beneath concrete and steel—vanished from the memory of men, but alive in the river's soul.

Floods occur when rivers exceed their boundaries, spilling water across the surrounding land. When rivers breach their banks, this flow is often more turbulent than the river's normal course, increasing its ability to erode the land and carry sediment. As they encounter obstacles like hills or buildings, floodwaters can experience abrupt changes in speed and height, known as *hydraulic jump*, which can substantially increase erosion in localized areas.

Despite their destructive potential, floods are nature's way of redistributing fertile soil. And the dilemma with flood control measures such as levees and dams is that they disrupt the very process that can build the delta. By containing the river, the natural deposition of sediment gets halted resulting in an accelerated land subsidence. This is becoming increasingly evident around the globe. For example, in the Mississippi delta in America, or in Egypt where the Aswan High Dam stopped the Nile's annual floods that had fertilized the delta with nutrient-rich silt ever since, causing coastal erosion and soil degradation. Likewise in Asia, a cascade of upstream dams in the mighty Mekong has reduced sediment flow to its delta in Vietnam, leading to saltwater intrusion. This is the paradox of controlling nature when the real question should have been about learning to live with it.

The River of Sorrows

In ancient China, around 2200 BCE, the Yellow River earned its mournful name: *River of Sorrows*. Year after year, it raged and swallowed entire villages beneath its torrents. To the eyes of mortal men, these floods felt like a punishment from an immortal divine. It was as though the land itself were paying the price for some past sins buried in its stones and soil. Fair or not, little did it matter, for this was a curse without an end.

The Emperor of the time, Yao, was desperate to protect his people. He summoned the sharpest minds the land had known. One after another, the empire's finest came forth; engineers with blueprints, strategists with grand visions. Some urged the building of huge walls to hold back the waters, others proposed reservoirs to catch and contain the floods. The emperor, driven by hope and the cries of his people alike, gave each one a chance...

And every single one of them failed. The walls crumbled. The reservoirs overflowed. And each year flood returned and it returned stronger than the last. With each such failure, the emperor's frustration grew further. The stakes were as high as life and death, literal and not just the metaphorical kind since those who failed to deliver were punished severely with many executed for their inability.

When all hopes had seemed to wash away, only then, did it feel the river gods had finally listened. From the floodplains of despair emerged a young man. He was the son of a high-ranking official who had also tried to control the floods. His father first met the failure and then, as expected, death. Now, the son stood before the emperor, ready to take on the same impossible task under the shadow of his father's fate. Unlike those before him who sought to fight the river, he asked to work with its flow, and not against it. The emperor, torn between curiosity and skepticism, gave him permission. Given that all else had failed, what choice did he have?

His mission began with a task as mammoth as the river itself: to map its every bend. On foot, he traveled across the empire to study all channels and tributaries of the Yellow River. It is said that his journey lasted over 13 years in which he climbed mountains to trace the river's source, waded through flooded plains, endured storms, hunger, cold—all to understand how the water moved, and where it surged.

According to the legend, during those 13 years, he passed by his own home on three occasions. Each time, he heard the voice of his family but never stopped. The first time, his wife was in labor. The sounds of childbirth filled the air yet he walked on. The second time, years later his son called out his name. By then, floods were at their peak and he could not afford to stop. The third time, more than a decade later, his son was standing at the door. Again, he moved on as his singular focus was to complete the task at hand.

His idea was to divide the water and give it more pathways to follow. It was only possible by digging diversion channels to direct the floodwaters into other rivers, in order to reduce the strain on the Yellow River. He also designed an extensive network of irrigation canals to carry the excess water across the fields that would feed the land instead of drowning it.

Such a colossal feat—unmatched in scale and unheard of before—demanded that regional rival tribes set aside their enmity and come together to agree, unanimously, on the map of channels. To make it work, he met rival tribal chiefs and local lords under the authority of Emperor Shun (who ruled after Yao). He told them about the plan and secured their cooperation in devising canal routes. It was a moment of rare unity, something unprecedented, and without which the entire system would have collapsed.

The daunting task of digging canals was started with the help of locals from every area. During all these times, he labored alongside common folk and dug with his own hands. At long last, the toil of years gave way to form and the great lattice of channels stood complete. And with it, a new chapter in the history of mankind was written. The following year, for the first time in living memory, the Yellow River did not bring misery. The floods that had tormented the land and its people for generations were finally subdued.

Moved by the harmony he brought between men and river, the emperor chose him as his successor. The man who had walked for thirteen years, who dug with his own hands, who passed by his family three times but never stopped eventually went on to rule ancient China for 45 years. More than 4,200 years later, he remains known to one and a half billion Chinese souls and some more beyond. We call him Da Yu, or *Yu the Great*.

He did not conquer the land to feed his ego like Alexander the Great. Nor did he butcher the men to build pyramid of skulls like Tamerlane. He conquered the river and helped the living survive. He was "great" by every measure that matters, or perhaps that *should* matter.

Rivers speak their own tongue. They carry within them the mystery of the Indus Valley Civilization. They narrate to us that no matter how advanced a society may become, it remains at the mercy of forces greater than itself. The Harappans placed their faith in the constancy of rivers, building their world on the belief that water would forever follow the same path. However, rivers, like time, are not fixed. They are always moving forward.

Flow

Doesn't that offer something worthy of reflection about the natural world? That nothing in it stays still—not time, not rivers, not landscapes, not even civilizations. Isn't that a hint for us? That we are meant to hold things loosely. The illusion of permanence, the urge to attach to people, places, or plans only invites sorrow. Why should we mourn what we know is destined to slip away? Would it not be wiser to live rightly within it, for as long as it remains? Like a good traveler who keeps on walking, grateful for the view, yet never expecting the road to remain the same.

Finally, what about those tiny droplets—the ones that make rivers what they are? The little droplet that rose from the sea, lifted by the sun's warmth, carried by the wind, fell to earth again, wandered through streams and rivers, nourished life, and left traces of itself wherever it went. And then one day it found its way back to the ocean.

Aren't we all like those droplets? No matter how far we travel, no matter the storms we face, despite our wanderings and despite our trials, we too are just passing through, part of something larger while finding our ways back. And in the end, we will complete the circle and return to where we all began... Back to the source... Back to where we all belong.

12

Flow of Life

Life flows...

Not in straight lines, nor in predictable patterns, but like water seeking its course. Sometimes swift, sometimes sluggish, at times crashing against jagged rocks, and at others pooling into stagnation before moving forward.

It is fluid, yet stiffens into rigidity. Although adjustable, it likes attaching to familiarity. But it never stops and neither should it. Only in motion does it stay pure and whole.

Nevertheless, unlike the rivers and winds, we often move not with clarity and purpose, but out of fear. Fear of missing out. Fear of losing an unknown race.

We run, without understanding where we are headed. We chase, without knowing what we are after. We see others rushing, and so we rush. We see someone accumulating, and so we gather. We see the rest applauded, and so we try to mold ourselves into what might earn us the same approval.

And in this blind pursuit, life spirals into chaos like a wild storm. We lose sight of why and how we ever set out in the first place. We end up exhausting ourselves not from the flow of life but from the overpowering need to impress upon those who are too consumed by their own races to care.

We treat life as a hundred-meter dash, when in reality, it is a marathon. An unforgiving stretch where endurance matters more than speed. And yet, somehow, we keep forgetting this.

We expect results overnight and want to race toward an imagined "there" as hurriedly as possible. Though we are rarely certain where that there actually is. We want it all, and we want it all now. Be it money without the wait, success without the struggle, or the peak without the climb. But there are no short cuts in life and nothing *lasting* ever happens in a flash.

In this marathon, million others are also running alongside us. Some are moving ahead, some falling behind. But tell me—if you saw someone sprinting past you, would you abandon your pace and start running frantically, just to catch them?

Do you know where they began? Do you know where they are going? Was this their first lap or tenth? Do you even know if they are running toward the finish line or just spinning in circles? You don't know anything about them.

And still, at the mere sight of someone moving faster, we begin to feel unsettled. It stirs a bitterness in us. In our haste to keep up, we make ourselves rush toward an unknown destination on an undefined timeline, only because someone else appeared to be ahead of us in some particular moment.

But what sense does that make?

Would a seasoned marathoner panic upon seeing a sprinter rush past? Would they throw away their strategy, burn their energy too soon, and risk never finishing? No. And not because they cannot run fast but because they understand the nature of their race.

So why don't we? Why do we measure our journey against someone else's path of which we know nothing about?

However, for most of our lives, we believe—or are led to believe—otherwise. In school, we compete with our classmates from a local town. We move to high school at the district level, and now the competition expands to students from other schools. In college, the rivals come from across the state. We go abroad, and we find ourselves competing on a global stage.

Do you see the pattern? Do you see the problem? Do you even remember the names of all your school classmates you ever 'competed' with? Can you recall every so-called 'rival' from high school or college?

The reality is there is no competitor in this marathon. There never was any. We are not here to outrun or outearn others. It *should* have always been about us—from the day we were born to the day we die. Each day, the goal should not be to run faster than other strangers, but to move better than how we moved yesterday... To seek a little progress... To think a little deeper... To act a little wiser.

Flow

A stream of water does not ask where the other streams are going. It does not imitate their flow neither does it seek applause for the trail it leaves in its wake. It moves to find its way to the ocean as that is its nature. And perhaps that is how we should flow too.

But who says progress will not be painful? Make no mistake, life is not without resistance. Obstacles will arise again and again to test the ability of our flow.

Have you ever seen a river quarrel with the rocks in its path? It does not stop to complain, nor does it wait for them to move. It flows around them, over them, and when needed, even through them. The wind does not stop if a mountain stands in its way. It shifts its path and continues. The mountain may be fixed, but the wind does not need to be.

So why do we stop when we should be flowing? How often do we resist the inevitable, refusing to bend? How often do we treat challenge as failure? As water turns when it must, and air bends when it must, so too must we loosen our grip on certainty and comfort.

A pebble sinks, whereas a ship floats. It has little to do with the water and even less to do with their sizes. It all comes down to their design.

We are no different either. The weight of life does not sink us. It is our design—how we carry what we carry—that determines whether we rise or drown. Ask yourself:

Am I built to stay afloat?
Do I recognize what lifts me?
What, or who, is my buoyancy?
Do I have the strength to release what pulls me under?

The things that keep us from sinking are not loud. It could be a kind word spoken just in time. It could be the presence of someone who stays when we have no strength left to ask. It could be a belief that refuses to abandon us, even when we have nearly let go of ourselves.

Also, look at a swimmer to see how he stays afloat. Now, as contrast, picture someone who cannot swim. The moment he will fall into the water, panic will take over. He would thrash his arms and kick his legs and he would splash wildly in every direction—grasping at everything and nothing at once.

Will he swim, or will he drown? And if he drowns, would he be justified in complaining: *I was moving. I was trying. I gave it everything all I had.*

Would you not tell him that haphazard movement only leads to exhaustion? That water does not acknowledge disorganized struggle? That it only responds to skill, and not to desperation?

Remind yourself of this too, especially in moments when life feels like an open sea and you are thrashing with all your might just to stay above the surface. Panic is instinctive. But panic does not keep anyone above water. Nor does blind, aimless struggle. Only when move-ment is deliberate, does the water hold the swimmer. And only when your energy stops scattering in every direction, would you stay afloat.

Water does not resist change. Under heat, it rises to steam. Under cold, it hardens to ice. However, whatever form water may take—ice, steam, or liquid—it always remains two atoms of hydrogen and one of oxygen.

Change as the situation demands but never lose what you are made of. Adapt, adjust, evolve, all the while staying true to your principles the way water stays true to its molecular structure.

It is a folly to assume that life flows like a laminar stream; smooth, predictable, untroubled. When, in fact, it is turbu-lent, full of chaos, and uncertain.

All of us are, by nature, creatures of structure. We tend to seek stability, or at least a sense of order. We go after a life shaped by predictability, underpinned by routines we can trust. But just when we begin to feel we have achieved it, the ground starts to shift and our plans falter. Yes, it stings badly and it drains us of hope.

As our private worlds shake, we look outward only to find the outside world in flames as well. And the ache does not stay personal anymore. Even hospitals are being bombed and reduced to dust. Innocent children lie buried beneath the rubble. Those who escape the blast are being starved by design.

We call ourselves children of God, or God's chosen people, don't we? But what kind of a chosen one chooses to slaughter the kids? What kind of a child murders other children? In our pride, if we refuse to see them as children of God—are they not even children of humans? In our prejudice, if we deny them chosenness or divinity—can we not, at least, grant them humanity?

In the anguish that follows, we are left with questions that rot in the throat. Questions that are too heavy to voice but at the same time too bitter to swallow. We dare not bring them to our lips, but we ask ourselves inside our hearts that where are the gods now? The world is burning and still there's no thunder from the skies. Why is there so much suffering and why does it so often fall on those who least deserve it? Is this what justice looks like where the innocent gets killed and the guilty go free? Why is the world in such disorder? Why so much chaos to begin with? And why must there be turbulence at all?

The truth about turbulence is that it may feel like pure disorder; a force that resists all structure, all reason. However deep within its wildness, there lies a hidden pattern. An elusive order that science, with all its tools, has yet to fully grasp. We do not know its form. We do not know its shape. We do not know what it looks like—but we do know, with certainty, that the order exists.

It may well be that the same is true of our chaotic world. The turbulence around us—wars, disasters, hunger, injustice—feels brutally random. It is thus easy in these moments to believe that all we have is disorder, and no one is watching. However what if that is only the surface. What if, like turbulence in a fluid, beneath the chaos lies a meaning? One beyond the limits of our perception? An order too complex for us to understand just yet? A pattern too intricate to be within our reach, at least for now.

The task at hand not to curse the chaos. Turbulence will rage as it always has, however, we must not let it swallow us whole. In the face of disorder, we must act with courage and hold fast to compassion. We must speak with unshakable clarity, even if ours is the only voice and even when silence feels safer.

We may not yet see the full pattern and perhaps that is not the point. What matters is that we can still choose to belong to the order that resides within the turbulence.

It captivated me for long that why do the principles of fluid mechanics align so intimately with human experience? Why do flow, turbulence, viscosity, and nearly all concepts and properties meant for fluids feel so familiar?

Then one day, a forgotten memory from my school years resurfaced that nearly 60% of the human body is fluid. Even more so, our brains are 75-80% fluid by volume. With every heartbeat, a river of blood flows through us. The air we breathe who knows may have once howled through a hurricane before finding its way to our lungs.

That's why we have within us the fury of floods and the serenity of streams. We show the rage of storms and the pleasantness of breeze.

We live *with* the fluids and we live *as* them.

To understand life is to understand its fluid nature. Because when all is said and done, that's precisely what we are—a living, breathing manifestation of fluid mechanics.

Yes, this was about science.

But more so, this was about us.

The café lights have dimmed, and the chairs are being stacked. We have had twelve cups of coffee together but now the *flow* of our conversation comes to rest.

Thank you for sitting with me all this while. Wherever you go, I hope you carry some of the ripples with you. And if you found value in this journey, I'd be truly grateful for an honest review on whichever platform you've traveled. It helps the work improve—and it may help others find the flow.

We never know what life holds for us next. For now, though, the cups are empty, the rain has cleared, and it's time to say good bye...

InkTangible
From Thoughts to Words